Rock Mechanics

Rock mechanics is a multidisciplinary subject that uses the principles of mechanical science to investigate rock engineering behavior. An important part of rock mechanics is laboratory studies of rock behavior. *Rock Mechanics: Theoretical Foundations of Laboratory Tests* is the first book that introduces the theoretical foundations of laboratory testing in an analytical manner.

It covers all main aspects, from the definition and classification of intact rock, sampling and specimen preparation, petrographic description and physical properties, and wave velocity and dynamic elastic constants, via durability and deterioration, hardness and abrasiveness, seepage and permeability, and expansion and swelling, to compressive strength and deformability, tensile strength and shear strength.

This book can be used as a textbook for undergraduate, master's and doctoral students in engineering geology, civil engineering, geotechnical engineering and mining engineering (rock mechanics). It will also appeal to professionals in consulting engineering companies and rock mechanics laboratories, and to geological, civil and mining engineers.

Rock Mechanics

Theoretical Foundations of Laboratory Tests

Davood Fereidooni

CRC Press
Taylor & Francis Group
Boca Raton London New York Leiden

CRC Press is an imprint of the
Taylor & Francis Group, an **informa** business
A BALKEMA BOOK

First published 2026
by CRC Press/Balkema
4 Park Square, Milton Park, Abingdon, Oxon, OX14 4RN

and by CRC Press/Balkema
2385 NW Executive Center Drive, Suite 320, Boca Raton FL 33431

CRC Press/Balkema is an imprint of the Taylor & Francis Group, an informa business

CRC Press is an imprint of Informa UK Limited

British Library Cataloguing-in-Publication Data
A catalogue record for this book is available from the British Library

ISBN: 9781041140412 (hbk)
ISBN: 9781041140924 (pbk)
ISBN: 9781003672920 (ebk)

DOI: 10.1201/9781003672920

Typeset in Times New Roman
by codeMantra

In the name of God

Contents

Preface

Rock mechanics is a multidisciplinary topic that uses the principles of mechanical science to investigate rock engineering behavior. A part of the rock mechanics science is related to investigating the behavior of rock mass in situ, which is called rock engineering. While another part of it studies the engineering and mechanical behavior of rocks in the laboratory, this part of the study is very important in engineering geology, geotechnical engineering and mining engineering. Therefore, having a knowledge of rock mechanics is very necessary for engineering geologists, geotechnical engineers and mining engineers. The main purpose of writing this book is to introduce the theoretical foundations of laboratory tests in an analytical way with the help of criticism and presenting the author's opinion and experiences, which is prepared in a simple and understandable way for the mentioned experts. It is hoped that it will open the way for those who are interested. This book was written by a university faculty member who has many records in the field of performing rock mechanics laboratory tests, conducting research projects and guiding student theses related to the topic and has significant records in teaching rock mechanics. The author sincerely thanks everyone who helped in the preparation of this work, as well as everyone who provided constructive feedback.

This book has 11 chapters: in Chapter 1, information about intact rock and its definition and classification is provided. Chapter 2 deals with sampling methods from the site and specimen preparation in the laboratory. In Chapter 3, the description of rock petrography and the method of determining the physical properties of intact rock are presented. In Chapter 4, the ultrasonic wave velocity and the dynamic elastic constants of intact rock are investigated. In Chapter 5, the subject of durability and deterioration of rock and its study methods are considered. In Chapter 6, the hardness and abrasiveness of intact rock are investigated and discussed. Rock permeability, measurement method and influencing factors are presented in Chapter 7. Chapter 8 deals with the analysis of rock expansion and swelling. In the ninth chapter, the knowledge of the compressive strength and deformation of intact rock is discussed. In Chapters 10 and 11, the tensile and shear strengths of intact rock, their measurement methods and factors affecting them have been presented.

About the author

Davood Fereidooni, a full-time faculty member at the School of Earth Sciences, Damghan University, Iran, is an accomplished academic and researcher with over two decades of experience in engineering geology and rock mechanics. He has also spent five years working in executive roles with a consulting engineer specializing in engineering geology and geotechnical engineering. Since 2014, he has been serving as a faculty member at Damghan University, primarily contributing to postgraduate education. He is currently the Head of Department and has been honored three times as Distinguished Researcher and once with an Educational Excellence award. Dr. Fereidooni holds a patent for a geotechnical testing apparatus and has authored four books, translated three others, published more than 50 peer-reviewed research articles in international journals, and presented over 40 conference papers. He has contributed to the successful completion of three national research projects and serves as an associate editor on the editorial boards of two reputable international journals. His research interests include engineering geology and rock mechanics in laboratory and field.

Chapter 1

Intact rock

Definition and classification

1.1 Introduction

Rocks are the constituent parts of the earth's solid crust, which are seen together in different forms. According to the definition, rock is a solid, natural, inorganic, heterogeneous and anisotropic material composed of one or several minerals. It should be noted that although the rocks are formed from the accumulation of minerals, this accumulation is not accidental and depends on the rock formation conditions and environment as well as the available primary materials for rock-forming. Therefore, these conditions can also be effective on the engineering characteristics of rocks (Fereidooni 2014). The application of rocks as engineering materials and carrying out construction operations on them or within them go back several centuries. The application of rocks includes foundation construction, rock slope creation, underground excavation, building materials and other such cases. The first recorded activities regarding the application of rocks include the construction of historical structures and monuments in Iran, Greece, Egypt, Iraq, India, China and so forth. The Anubanini petroglyphs (28th century BC), Persepolis (5th century BC), Ganjnameh and Bistoon inscriptions (4th century BC), Giza pyramids (26th century BC), Abul-Hail temple (12th century BC), Hanging Gardens of Babylon (6th century BC), Temple of Olympian Zeus (6th century BC) and Parthenon Temple (5th century BC) are among the most important cases of using rocks in the construction of historical structures and buildings (Figure 1.1).

In the construction of structures and engineering projects, we are always faced with a large area of rock, which is called rock mass. The rock mass may have dimensions from several centimeters to several kilometers and usually includes many discontinuities of different types. The presence of discontinuities such as faults, joints, fractures, cracks, bedding planes and lamination in the rock mass turns it into blocks of healthy and undisturbed rocks, which are called intact rocks. In other words, the intact rock refers to unbroken blocks between the rock mass discontinuities. The size of these blocks may vary from a few millimeters to several meters. Therefore, usually, the rock mass can be investigated on a large or field scale and intact rock on a small or laboratory scale (Fereidooni 2004; Fereidooni and Khodadadzadeh 2014). The engineering properties of intact rock are controlled by the physical and mechanical properties of its constituent minerals and their connection to each other. The parameters that may be used in the description and evaluation of intact rock engineering properties include lithological name and characteristics, color, texture, grain size, density, porosity, hardness, strength and deformability. The intact rock is described and classified from both geological and engineering points of view, which are discussed in this chapter.

DOI: 10.1201/9781003672920-1

Figure 1.1 Historical cases of rock application: (a) Persepolis, (b) Anubanini petroglyphs, (c) Ganjnameh inscriptions, (d) Parthenon temple and (e) Temple of Olympian Zeus.

1.2 Geological classification of rocks

Geologists classify the different rocks in the earth's crust into three main categories, including igneous rocks, sedimentary rocks and metamorphic rocks. These three groups have always been changing and evolving throughout geological time and have constantly transformed into each other. The process of transforming rocks into each other is called the rock cycle. As previously mentioned, intact rock is classified in terms of geology or engineering. In the first case, the mineralogical content of the rock, its texture and any changes that have occurred since the formation of the rock have great importance. Although the geological classification of intact rock usually has a genetic basis, it provides little information regarding the engineering behavior of the rock, because, for example, intact rocks of the same geological class may show a large scatter in the value of strength, deformability and overall engineering properties, and this difference can reach several times. Therefore, carrying out engineering classifications of intact rock, which are mostly related to the engineering properties of rocks and the parameters affecting them, is very necessary and important in rock mechanics science.

1.2.1 Igneous rocks

The igneous rocks are obtained from the crystallization of molten and mobile materials within the earth's crust, that is, magma. In other words, magma is the primary material that forms the igneous rocks. The lexical root of magma is a Greek word meaning paste, and according to the definition of magma, it refers to molten and silicate materials that contain a large amount of gas and volatile substances in solution state in its composition. If the gas and volatile substances escape from the magma, lava will be formed. This action happens when the pressure is removed from the magma, or in other words, the magma reaches the earth's surface. The freezing of magma and its transformation into the igneous rocks is called crystallization. In fact, crystallization is the change of magma state from molten state to solid state that happened in one of the crystallization systems. The crystallization of magma may take place on the surface or in the interior parts of the earth's crust. Therefore, depending on where the magma crystallizes, the igneous rocks are classified into two main groups, namely, intrusive and extrusive igneous rocks. The first type is those igneous rocks that are formed from the crystallization of magma within the earth's crust. These rocks are mostly composed of coarse-grain crystals. The second type is formed from the crystallization of lava on the earth's surface and is mostly microcrystalline. In addition, the igneous rocks are classified based on other characteristics such as texture, quartz percentage, color, density and chemical composition. Table 1.1 presents the most important types of igneous rocks and their characteristics.

Table 1.1 The most important igneous rocks and their characteristics (Zhang 2006)

Group/properties	*Granitic (acidic)*	*Andesitic (intermediate)*	*Basic (basic and mafic)*	*Ultrabasic (ultrabasic)*
Intrusive rocks (coarse-grained)	Granite	Diorite	Gabbro	Peridotite
Extrusive rocks (fine-grained)	Rhyolite	Andesite	Basalt	Not exist
Quartz percentage	>65	65–55	55–45	<45
Primary minerals	Quartz, orthoclase and sodium plagioclase	Amphibole, plagioclase and biotite	Calcium plagioclase and Pyroxene	Olivine and pyroxene
Secondary minerals	Muscovite, biotite and amphibole	Pyroxene	Olivine and amphibole	Calcium plagioclase
Color	Light ------------------------------▶ Dark			

1.2.2 Sedimentary rocks

The sedimentary rocks are a collection of solidified sedimentary materials that have accumulated at low depths, under low pressure and temperature conditions, and have become compact and hard. These rocks make up about 5% of the earth's crust volume and cover about 75% of the earth's surface. On the other hand, igneous and metamorphic rocks include about 95% of the volume of the earth's crust and cover about 25% of its surface. Therefore, it can be stated that the sedimentary rocks are formed at shallow depths. In the formation of sedimentary rocks, various processes are involved from the moment of sediment formation to their transformation into sedimentary rocks. These processes include physical, biological and chemical reactions as well as all the factors that cause the transformation of the sediments into sedimentary rocks. These factors include weathering, transportation of materials resulting from the weathering (sediments), deposition of sediments in sedimentation basins, lithification and uplift. The sedimentary rocks are classified into three groups, namely, clastic or detrital rocks, chemical and biochemical rocks and pyroclastic rocks.

1.2.2.1 Clastic rocks

These rocks are composed of clastic particles or debris. These particles are the result of the erosion of parent rocks that are located outside the sedimentation basin. The clastic particles make up about 60%–80% of the total sediment particles and include rock fragments, quartz, feldspars, clay minerals, mica and heavy minerals. The clastic particles that make up the sedimentary rocks have different sizes, and accordingly, the clastic rocks can be classified. Table 1.2 presents a classification of clastic particles based on size and types of sedimentary rocks formed from these particles. According to the size of constituent particles, the clastic rocks are classified into three groups including (i) coarse-grained rocks or conglomerates and breccias, (ii) medium-grained rocks or sandstones and (iii) fine-grained rocks or mudstones.

Table 1.2 Classification of the clastic particles based on their size and the composed sedimentary rocks

Particle name		*Particle size (mm)*	*Sedimentary rock name*	
Gravel	Boulder	>256	Conglomerate and breccia	
	Cobble	256–128		
	Pebble	128–16		
	Granule	16–2		
Sand	Very coarse sand	2–1	Sandstone	
	Coarse sand	1–0.5		
	Intermediate sand	0.5–0.25		
	Fine sand	0.25–0.125		
	Very fine sand	0.125–0.0625		
Silt	Coarse silt	0.0625–0.031	Siltstone	Mudstone (shale)
	Intermediate silt	0.031–0.0156		
	Fine silt	0.0156–0.0078		
	Very fine silt	0.0078–0.0039		
Clay		<0.0039	Claystone	

1.2.2.2 Chemical and biochemical rocks

The constituent particles of these types of rocks originate from the sedimentary basin and are chemically or biochemically deposited. In terms of composition, chemical and biochemical rocks are classified into seven general groups of carbonate rocks (limestone and dolomite or dolestone), chert, evaporite rocks, phosphates, iron-bearing rocks, manganese rocks and carbonaceous rocks. Table 1.3 shows the classification of the most important types of chemical and biochemical sedimentary rocks.

1.2.2.3 Pyroclastic rocks

This category of rocks is formed by the eruption of volcanoes and the sedimentation of materials resulting from this eruption in water or on the ground surface. Some geologists classify these rocks in the group of igneous rocks and others in the group of sedimentary rocks. The pyroclastic rocks are named based on the size and type of constituent particles and their origin. Table 1.4 summarizes a classification for these types of rocks based on the size and type of their constituent particles.

1.2.3 Metamorphic rocks

The metamorphic rocks are the third group of rocks that have been classified and studied by geologists. These rocks are formed from the deformation of previously mentioned rocks due to

Table 1.3 Classification of the most important types of chemical and biochemical sedimentary rocks (Fereidooni 2014)

Rock type	*Group*	*Rock name*	*Description*	*Main composing minerals*
Chemical	Iron	Iron ore	Limestone or claystone containing iron (or evaporite)	Calcite, iron oxide
	Carbonate, silicate and evaporite	Dolomite and limestone	Evaporite or replacement limestone, fine-grained	Dolomite and calcite
Biochemical	Carbonate	Limestone	Contains fossils, coarse-grained or fine-grained	Calcite

Table 1.4 Classification of the pyroclastic rocks (Fereidooni 2014)

Particle size (mm)	*Particle name*	*Sediment*	*Pyroclastic rocks*
<64	Piece or bomb	Agglomerate, layers of bombs or fragments, tephra containing fragments or bombs	Agglomerate and pyroclastic breccia
64–2	Lapillus	Layers of lapilli or lapilli tephra	Lapilli
2–1/16	Coarse-grained ash	Coarse-grained ash	Coarse-grained tuff
<1/16	Fine-grained ash (dust)	Fine-grained ash or dust	Fine-grained tuff or dusty tuff

Table 1.5 Classification of the most important types of metamorphic rocks

Group	*Rock*	*Description*	*Main composing minerals*
Massive	Hornfels	Very fine-grained	Quartz
	Quartzite	Fine-grained	Quartz
	Marble	Fine-grained to coarse-grained	Calcite or dolomite
Foliated	Slate	Very fine-grained, laminated	Kaolinite, mica
	Phyllite	Soft and foliated	Mica, kaolinite
	Schist	Coarse-grained	Feldspar, quartz, mica
	Gneiss	Metamorphized from granite	Hornblende

changes in physical (pressure and temperature) or chemical conditions in a solid state. In fact, the phenomenon of metamorphism is interpreted as the disappearance of one or a set of minerals and the appearance of other minerals. These changes may occur on the sedimentary rocks that were formed in the conditions of the earth's surface on the igneous rocks that crystallized from magma or even on previous metamorphic rocks. In the last case, the metamorphosis conditions of the previous rock change, and this phenomenon is associated with the emergence of one or a set of new minerals. Therefore, the metamorphism process is the response that each rock expresses to changes in the chemical or physical environment. This response is manifested in the form of recrystallization of old minerals or the emergence of new minerals and destruction of some previous minerals.

As previously stated, the sedimentary rocks are formed in the conditions of the earth's surface during the process of diagenesis at a temperature less than 150°C, and the igneous rocks are formed at high temperatures, between 650°C and 1200°C, as a result of melting. In terms of pressure and temperature conditions of formation, the metamorphic rocks are an intermediate group between the two mentioned rock groups. Therefore, it can be stated that diagenesis is the lower limit of the metamorphism process, and melting is considered the upper limit of this process. Table 1.5 shows the classification of the most important types of metamorphic rocks.

1.3 Engineering classification of intact rock

Engineering classifications of intact rock are based on the strength or deformation characteristics of rocks. As an example, Table 1.6 shows the International Society of Rock Mechanics (ISRM 1978) classification system for intact rock. In this classification, the intact rock may range from very weak to very strong depending on uniaxial or unconfined compressive strength or field identifications. Based on laboratory measurements of the strength and deformation properties of rocks, Deere and Miller (1966) presented an intact rock classification system using unconfined compressive strength and tangential modulus of elasticity at 50% of ultimate rock strength. Figure 1.2 shows the engineering classification of the igneous, sedimentary and metamorphic rocks based on the classification proposed by Deere and Miller (1966). In this figure, the modulus ratio is the ratio of the modulus of elasticity to the unconfined compressive strength of the rock. In this classification, there are 15 groups of intact rock, and each category is identified by a two-letter name. The first letter in the category name represents the unconfined compressive strength, and the second letter represents the modulus ratio of the rock. Therefore, based on this classification, an intact rock may be placed in the category of AH, AM, AL, BH, BM, BL and so forth, which have different characteristics in terms of strength and deformation. Rock elastic properties can be omitted in practical rock classification, as the modulus of elasticity determined in the laboratory is rarely what is needed for rock engineering analysis in the field (Fereidooni et al. 2015).

Table 1.6 Engineering classification of rock by strength (ISRM 1978)

Grade	*Classification*	*Field identification*	*Unconfined compressive strength (MPa)*	*Point load index (MPa)*	*Examples*
R0	Extremely weak	Indented by thumbnail	<1	[a]	Stiff fault gouge
R1	Very weak	Crumbles under firm blows of geological hammer; can be peeled with a pocket knife	1–5	[a]	Highly weathered or altered rock, shale
R2	Weak	Can be peeled with a pocket knife with difficulty; shallow indentations made by a firm blow with the point of a geological hammer	5–25	[a]	Chalk, claystone, potash, marl, siltstone, shale, rock salt
R3	Medium strong	Cannot be scraped or peeled with a pocket knife; the specimen can be fractured with a single firm blow of geological hammer	25–50	1–2	Concrete, phyllite, schist, siltstone
R4	Strong	Specimen requires more than one blow of the geological hammer to fracture	50–100	2–4	Limestone, marble, sandstone, schist
R5	Very strong	Specimen requires many blows of geological hammer to fracture	100–250	4–10	Amphibolite, sandstone, basalt, gabbro, gneiss, granodiorite, peridotite, rhyolite, tiff
R6	Extremely strong	The specimen can only be chipped with the geological hammer	>250	>10	Fresh basalt, chert, diabase, gneiss, granite, quartzite

[a] Point load tests on rocks with unconfined compressive strength below 25 MPa are likely to yield highly ambiguous results.

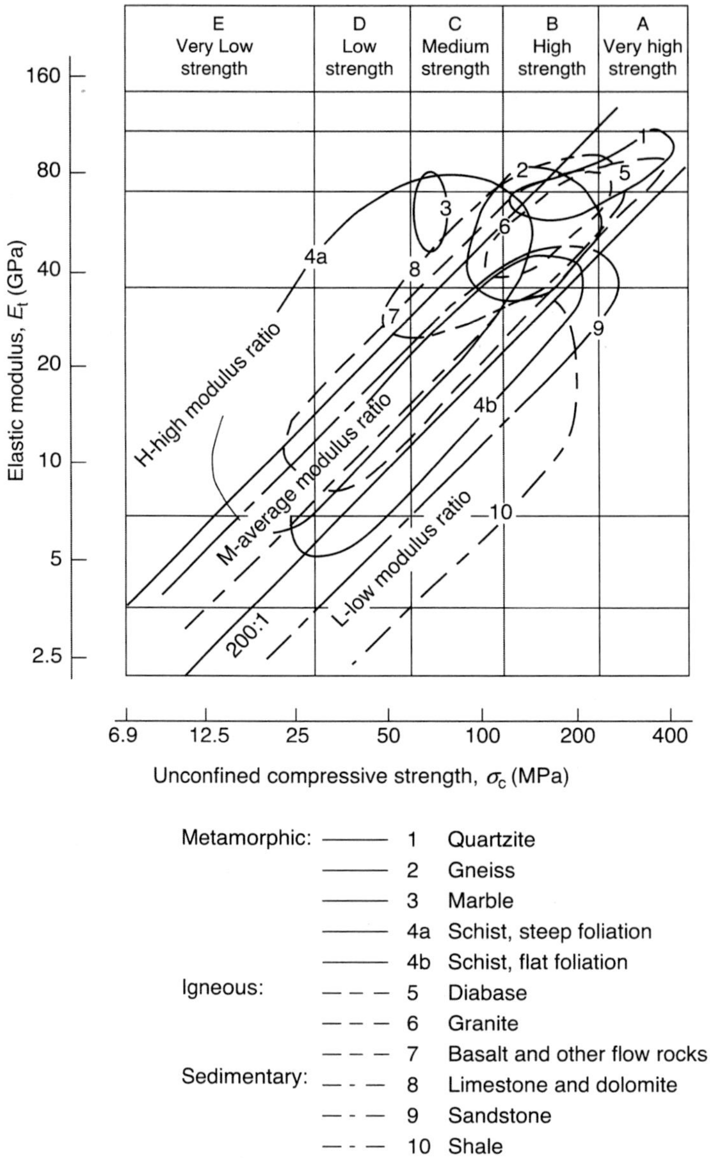

Figure 1.2 Engineering classification of intact rocks (E_t is the tangent modulus at 50% ultimate strength) (Deere and Miller 1966).

References

Deere, D.U., Miller, R.P., 1966. Engineering classification and index properties for intact rock. Tech. Rep. No. AFWL-TR-65-16, Air Force Weapons Lab, Kirtland Air Force Base, Albuquerque, NM.

Fereidooni, D., 2004. Investigating the stability of rock slopes along the Ganjnameh-Shahrestaneh road. Master's Thesis in Engineering Geology, Faculty of Science, Bu-Ali Sina University, 192 pp.

Fereidooni, D., 2014. *General and Engineering Geology*. Second edition, Daneshjoo Publications, Damghan, 450 pp.

Fereidooni, D., Khanlari, G.R., Heidari, M., 2015. Assessment of a modified rock mass classification system for rock slope stability analysis in the Q-system, *Earth Sciences Research Journal*, 19(2), 1–11.

Fereidooni, D., Khodadadzadeh, M., 2014. Evaluation of engineering characteristics of rock mass of Darband dam construction using DMR classification and kinematic analysis, Northeast Iran. *Proceedings of the Second National Conference on Soil Mechanics and Foundation Engineering*, Qom University of Technology, October 22 and 23, 2014, Qom, Iran.

ISRM, 1978. Suggested methods for the quantitative description of discontinuities in rock masses. International Society for Rock Mechanics, Commission on Standardization of Laboratory and Field Tests. *International Journal of Rock Mechanics and Mining Sciences & Geomechanics Abstracts*, 15, 319–368.

Zhang, L., 2006. *Engineering Properties of Rocks.* Geo-Engineering Book Series. Elsevier, Amsterdam, Vol. 4, pp. 1–290.

Chapter 2

Sampling and specimen preparation

2.1 Introduction

Sampling means choosing the best representative from the statistical population, so that the characteristics of that statistical population are preserved in the selected sample. In rock mechanics, the statistical population is rock mass, and one or more intact rock samples should be selected from it, so that its engineering characteristics are preserved in the selected samples. Basic sampling requires its methods, the expertise and experience of the sampler, appropriate tools and specific sampling time. By choosing the correct sampling method, the experienced person should choose a sample in which there is no weathering, joints and fractures, heterogeneity and other misleading parameters, and the selected sample is the best representative of the constituents of the rock mass. Sampling tools differ in surface and depth, which will be explained in the following sections. In terms of time, it is better to take samples during the day and under direct sunlight, because the characteristics of rocks can be recognized better at this time. During sampling, with the help of accurate visual observations and the use of a geological hammer, it is possible to ensure weathering conditions and the absence of joints and discontinuities in the sample. When hitting the surface of the sample with a hammer, if the treble sound is heard, the rock is free of weathering and there are no joints or discontinuities in the sample. But if the sound is bass, it indicates the effect of weathering and the presence of joints and discontinuities in it. It is very necessary to avoid sampling from weathered areas, debris and rocks whose characteristics have changed. Sampling of the rock mass can be done both from the surface and from the depth, each of which requires its methods and tools. The sampling of surface areas and deep areas are different from each other. For this reason, in the following, the principles of sampling from surface areas and the depth of the ground and their steps will be discussed.

2.2 Surface sampling

Sampling from the surface of rock mass is done when the rock mass is exposed on the ground surface and constructing surface structures such as slopes, surface foundations and roads are intended, or the purpose is to prepare building materials such as building stones and aggregates from the rock mass. Surface sampling from the rock mass always has its difficulties, because usually the rocks exposed on the earth's crust surface are weathered, which affects the selection of high-quality samples. Various tools and equipment are used in surface sampling of rock mass, such as geological hammer, masonry chisel and hammer, sledgehammer, crowbar and portable coring device. Depending on the amount of sample required, cores, pieces or blocks of suitable size and dimensions should be taken from rock mass, provided that these samples have the

DOI: 10.1201/9781003672920-2

characteristics of the rock mass and are not prepared from the weathered parts and discontinuity locations (Fereidooni 2023). To obtain a sample from the rock mass surface, there are various methods, the most important of which include random sampling, window sampling, simple linear sampling and cross or radial linear sampling.

2.2.1 *Random sampling*

In the random sampling, samples are randomly selected from different places of rock mass and taken by sampling tools. For example, in Figure 2.1a, locations for surface sampling of a roadside trench are shown. In this method, the sampler person has the freedom of action to choose the best sample from the rock mass because he/she has the authority to choose any location for sampling operations. The random sampling can be used to sample the surface of all rock mass types, including layered and laminated rock masses, or rock masses with a homogeneous and uniform surface (Fereidooni and Askari 2018).

2.2.2 *Window sampling*

In the window sampling, first, a window, or a hexagonal (honeycomb), square or triangular network is drawn on the rock mass surface. Then, one or more samples are taken from each window. The sampling place can be chosen in the center of the polygons or at their corners. In this method, it is possible that the sampling location is not located in a suitable place on the rock mass surface; for this reason, the freedom of action of the sampler person is, to some extent, denied. At this time, the sampler person should take the appropriate sample from the closest and best point to the sampling location with the help of his/her experience. This sampling method is also used for all types of rock masses, but it is limited to layered and laminated rock masses (Figure 2.1b).

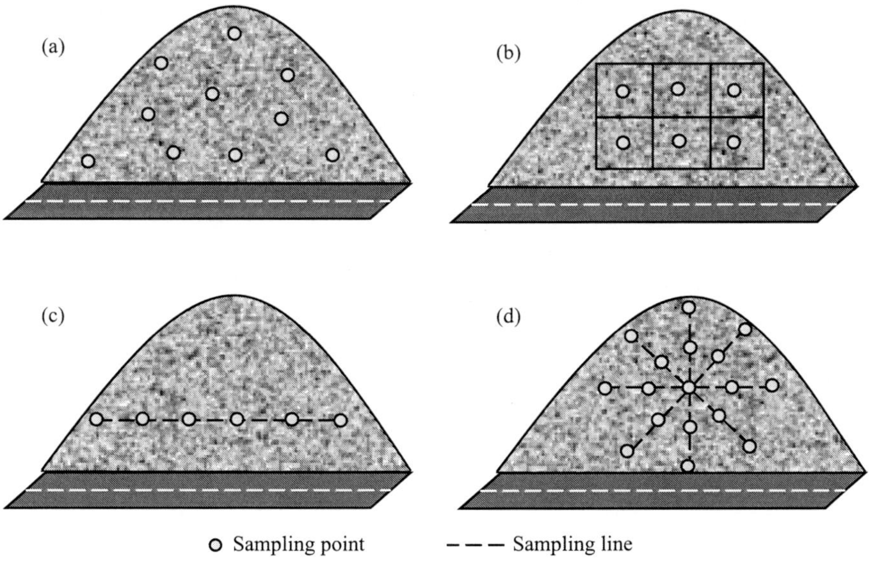

Figure 2.1 Different methods for sampling from the surface of the rock mass composing a road trench: (a) random sampling, (b) window sampling, (c) simple linear sampling and (d) cross linear sampling.

2.2.3 *Simple linear sampling*

In the simple linear sampling, first, a sampling line is selected with a suitable orientation on the rock mass surface. In rock masses with a uniform and homogeneous surface, the sampling line can be chosen in any direction, but in layered and laminated rock masses, it is better that the sampling line be selected perpendicular to the layering or laminating of the rock mass. Then, on the sampling line, samples are taken from the rock mass at equal intervals. If the sampling places correspond to the discontinuities of the rock mass or areas with severe weathering, the sampler person should choose a help point next to the main point to carry out the sampling operation (Figure 2.1c).

2.2.4 *Cross linear sampling*

In the cross-linear, or radial sampling, several sampling lines are selected in different directions to cover almost the entire rock mass surface. Then, on the sampling lines, samples are taken from the rock mass at equal distances. In this method, the sampling places may not be located in a suitable place, so that the sampler person can choose help points to carry out the sampling operation. Sampling using the cross or radial linear sampling in layered and laminated rock mass has many limitations, and for this reason, it is better to use it for rock mass with a uniform and homogeneous surface (Figure 2.1d).

2.3 Sampling from the depth

According to the sampling method and tool, different sampling methods from the rock mass depth can be divided into three general groups, including (i) block sampling, (ii) sampling by drilling and (iii) Integral sampling (Stölben and Eitner 2004). A combination of the mentioned sampling methods is possible and sometimes required according to the geological conditions of the area. In terms of geometric shape, rock samples taken from depth have different types, including (i) rock core (complete and incomplete), (ii) cut sample and crushed stone and (iii) block samples. In each of the sampling methods that are associated with coring, the quality of rock recovery is obtained by determining the following three parameters:

1. Total core recovery index (TCR): It is the total length of recovered solid and damaged core, which is expressed as a percentage of the length of the run or drilling stage.
2. Rock quality designation (RQD): It is the total length of all core pieces with at least one full diameter that is 100 mm or more and is measured between the natural fractures of rock. The rock quality designation is measured along the axial line of the core and calculated as a percentage of the length of the run or drilling stage (Khodadad-Zadeh et al. 2022).
3. Solid core recovery index (SCR): It is the length of core recovered as solid cylinders, which is calculated as a percentage of the run or drilling stage length.

After the recovery of the core barrels to the ground surface, the core recovery ratio should be determined. In cases where the core samples are removed from the sampler and placed in the core box, the information on the samples must be recorded. If core box storage racks are used, it should be decided in advance where and when the box will be opened to examine the cores. Core losses in the box must be filled with a dummy object. Also, the direction of drilling should be indicated on the core boxes or samples with an arrow, and the depth of the cores should also be specified. Depending on the best obtainable rock sample quality and ground conditions, there are three types of rock sampling methods:

- Grade A sampling: In this case, it is intended to obtain samples in which no or only minor disturbance of the rock structure occurred during the sampling process. In this case, the resistance and deformation properties, water content, density, porosity and permeability of the rock sample correspond to the in-situ values, and no change has occurred in the chemical composition of the rock mass. In this case, some unforeseen conditions, such as different geological conditions, can lead to a decrease in sample quality.
- Grade B sampling: In grade B sampling methods, the obtained samples must have all the constituent components of in-situ rock mass in their original proportions, and the rock pieces must have maintained their resistance and deformation properties, water content, density and porosity. By using grade B sampling, the general arrangement of discontinuities in the rock mass can be identified. In this case, the structure of the rock mass has been disturbed, and as a result, the resistance and deformation properties, water content, density, porosity and permeability of the rock mass itself have been disturbed. In this case, some unforeseen conditions, such as different geological conditions, can lead to a decrease in sample quality.
- Grade C sampling: In this case, the structure of rock mass and its discontinuities are completely changed and the rock materials may be crushed. Even some changes in the chemical composition of the rock may occur, but the rock type, matrix, texture and construction can be identified.

2.3.1 *Block sampling*

In the block sampling, samples are obtained from a test pit, well or borehole using special samplers and cutting. This sampling method is usually considered a grade A sampling method.

2.3.2 *Sampling by drilling*

During sampling by drilling from rock mass, the following points should be kept in mind:

- The drilling method and equipment should be selected as a function of the required sampling grade and geological and hydrogeological conditions.
- The drilling flushing medium (water passage necessary for drilling) should be selected in such a way that it meets the requirements of the research, and if necessary, suitable additives can be added to this medium. Hydrogeological requirements should be taken into account when selecting the flushing medium.
- In soft rocks, only double- or triple-tube core barrels should be used.
- The type of bit must be selected for efficient drilling and cutting of the rock type.
- Cementing and grout injection can be necessary to stabilize the borehole wall or sampling when passing through the crushed rock zone.
- The inclination of boreholes should be determined, including the maximum acceptable deviation, taking into account the investigation aims and the ground conditions.

2.3.2.1 *Sampling by rotary dry core drilling*

In the sampling by rotary dry core drilling, a drill rod system, with a bit attached to its lower end and rotated by the drill rig through the drill string, enters the rock mass. In this operation, a cylindrical core is obtained in the core tube. In this method, the sampling tool is a single tube with a borehole diameter of 70–200 mm. This sampling method can be used to recover core samples in soft, erodible and water-sensitive rocks, and it is less suitable for rocks with moderate to high hardness. To avoid overheating of the bit, the drilling run should not be more than 0.5 m. The sampling by rotary dry core drilling is a grade B sampling method (Table 2.1).

2.3.2.2 *Sampling by rotary core drilling*

In sampling by wet rotary drilling, a core tube equipped with a bit is screwed to the end by the drilling rig and with the help of the drilling string is entered into the rock mass. This operation causes the core to be taken in the core tube. In this method, the sampling tube can be a single, double or triple tube, and the diameter of the drilled borehole varies between 70 and 200 mm. In this method, a flushing medium is usually used. The single-tube core barrel consists of a coring tube, at the lower end of which a bit is installed and the upper end of which is connected to the drill rod. A core lifter can be installed between the bit and the core tube or directly within the bit. The flushing medium (water passage) is between the inner diameter of the coring tube and the recovered core within it and continuously washes the entire length of the recovered core. The double-tube core barrel consists of two concentric tubes and a bearing in the head of the core tube, which allows the inner tube to remain stationary, while the outer tube and bit are rotated by the drill string. A core lifter is installed between the bit and the core tube or directly within the bit. The flushing medium passes through the space between the inner and outer tubes, thus protecting the recovered sample from erosion and washout. The triple-tube core barrel system is similar to the double-tube core barrel, except that another tube is installed within the inner tube. Both double- and triple-tube core barrels are extensions of the inner tube in the direction of the bit and are suitable for use in very soft formations. With this method, samples can be obtained in the form of cores or cuttings. The single-tube core-barrel only provides the possibility of obtaining cores in resistant formations, while double- and triple-tube core barrels can be used in all types of rock formations (Fereidooni and Khajoond 2019). All core barrels can be fitted with plastic liners within the inner tube to assist core recovery and protection. Sampling by the rotary core drilling with single- and double-tube core barrels is usually a grade B sampling method, while sampling with the triple-tube core barrel is usually classified as grade A sampling (Table 2.1).

2.3.2.3 *Sampling by wireline core drilling*

In the sampling by wireline core drilling, a double- or triple-tube core barrel, with a bit fitted to the lower end, is rotated and fed into the rock to be drilled by the drill rig via the wireline drill rods. In this case, a core sample is taken within the inner tube. In this method, the diameter of the borehole varies from 70 to 180 mm. After completing a drilling run, the inner tube containing the sample is withdrawn with the help of the drilling rods using a wireline cable and winch. During this process, the drill bit, outer tube and drill rods remain at the bottom of the borehole. The sampling by wireline core drilling is a grade A sampling method (Table 2.1).

2.3.2.4 *Sampling of cuttings by rotary open hole drilling*

In the sampling by rotary open hole drilling a rock roller, drag or button bit is rotated and fed into the rock to generate cuttings. These cuttings are raised to the surface by the velocity of the flushing medium and collected or sampled at the borehole mouth. The drilled borehole diameter

Table 2.1 Features of rock drilling sampling methods (Stölben and Eitner 2004)

Column	*1*	*2*	*3*	*4*	*5*	*6*	*7*	*8*	*9*	*10*
Line	*Drilling method*			*Equipment*		*Drilling method less suitable for*[a]	*Samples*		*Achievable Sampling category*[b]	*Remarks*
	Flushing medium	*Extraction of sample by*	*Designation*	*Sampling tool*	*Guideline for borehole diameter range*[a] *mm*		*Cores*[a]	*Cuttings*		
1	No	Drilling tool attached to drill rods	Rotary dry core drilling	Single-tube core barrel	70–200	Rock of medium to high hardness	Soft, erodable, water-sensitive rock; short core runs	None	B (A)	To prevent overheating of the bit, core runs should not exceed 0.5 m
2	Yes	Drilling tool attached to drill rods	Rotary core drilling	Single-tube core barrel	70–200	Rock of medium to high hardness	Jointed, soft rock	Sieve residue and suspended matter	B (A)	Flushing medium can cause disturbance of core material
3	Yes	Drilling tool attached to drill rods	Rotary core drilling	Double-tube core barrel	70–200	Erodable, water-sensitive rock	All types of rock	Sieve residue and suspended matter	A (B)	—

(*Continued*)

Table 2.1 (Continued)

Column	*1*	*2*	*3*	*4*	*5*	*6*	*7*	*8*	*9*	*10*
Line	*Drilling method*			*Equipment*		*Drilling method less suitable for*[a]	*Samples*		*Achievable Sampling category*[b]	*Remarks*
	Flushing medium	*Extraction of sample by*	*Designation*	*Sampling tool*	*Guideline for borehole diameter range*[a] *mm*		*Cores*[a]	*Cuttings*		
4	Yes	Drilling tool attached to drill rods	Rotary core drilling	Triple-tube core barrel	70–200	—	All types of rock	Sieve residue and suspended matter	A	—
5	Yes	Drilling tool attached to drill rods, with wireline extractable inner barrel	Wireline core drilling	Wireline core barrel, or triple-tube core barrel	70–180	—	All types of rock	Sieve residue and suspended matter	A	—
6	Yes	Drilling tool attached to drill rods	Open hole drilling	Solid bit, roller bit, DTTH	50–350	—	None	Sieve residue and suspended matter	C	—

[a] Guideline values considering the possible use of a casing.
[b] The sampling categories given in parentheses can only be achieved in particularly favorable or unfavorable ground conditions, which shall be explained in such cases.
[c] In some crystalline rocks, a minimum borehole diameter of 30 mm may be sufficient for the identification and description of rock.
Note: The sample diameter is smaller for the same borehole diameter when a triple-tube core barrel is used, instead of a single-tube core barrel.

DTTH, down-the-hole-hammer.

in this method is usually between 70 and 311 mm. In this method, the core is not obtained and only disturbed cuttings are produced, and therefore, it is considered grade C sampling method (Table 2.1).

2.3.3 *Integral sampling*

In the integral or continuous sampling, complete, oriented and intact core samples can be taken to preserve the characteristics of rock mass (without disturbance caused by drilling) and initial conditions of natural discontinuities and their orientation. In this method, first, a borehole with a diameter of at least 25 mm is drilled in the rock mass and a perforated central tube is placed in this borehole. Then, this tube is filled with suitable connecting materials such as cement without pressure to connect the parts with penetration of the cement into the discontinuities of the rock mass. Finally, after the required setting time of the connecting material, a borehole is drilled coaxially with the first borehole and with a minimum diameter of 100 mm, and the core is taken out. It should be noted that the method of drilling for the second borehole must be following the criteria of grade B sampling method.

2.4 Carrying the sample to the laboratory

Before transporting samples to the laboratory, they should be properly packaged and labeled. It is better to name the sample in the form of an abbreviated letter of the sampling location name, together with a suitable number. Considering that the shape and size of samples are different, a suitable packaging should be considered for them. In this regard, core samples are usually placed inside a core box. Small samples and fragment or crashed samples may be placed inside special plastic nylons or wax-coated cloth to preserve their natural moisture. The samples should be transported by a suitable car or truck to prevent pressure or factors that cause changes in their characteristics and disturbances in their structure during transportation to the laboratory. Loading the samples into the truck has its own rules. Also, when unloading the samples and storing them in the warehouse or laboratory, it should be done in such a way that the least damage is done to them. Knowledge, experience and responsibility of people are very important when carrying samples to the laboratory.

2.5 Specimen preparation in the laboratory

The specimen preparation in the laboratory is done according to the requirements of tests and based on the methods proposed by ASTM (1996, 2001) and ISRM (2007). In this regard, various laboratory tests need the specimens without or with regular geometric shape. The specimens without regular geometric shapes are those specimens like lump-shaped specimens that are used in some tests. These specimens can be prepared in the laboratory using a geological hammer and sledgehammer. The specimens with a regular geometric shape include cylindrical, disc-shaped, cubic or rectangular cube specimens. The cylindrical specimens may have been obtained from drilling operations, in which case their heads and bottoms are cut in the laboratory. If a cylindrical-shaped specimen is to be taken from the blocks transferred to the laboratory, it must first be prepared with appropriate diameter cores with a laboratory coring machine. Then, the head and bottom of the specimens should be cut with the laboratory rock-cutting machine at the appropriate length for the test. In general, NX cores are prepared with a diameter of approximately 54 mm. According to the standards, a rock-cutting machine can be used to prepare

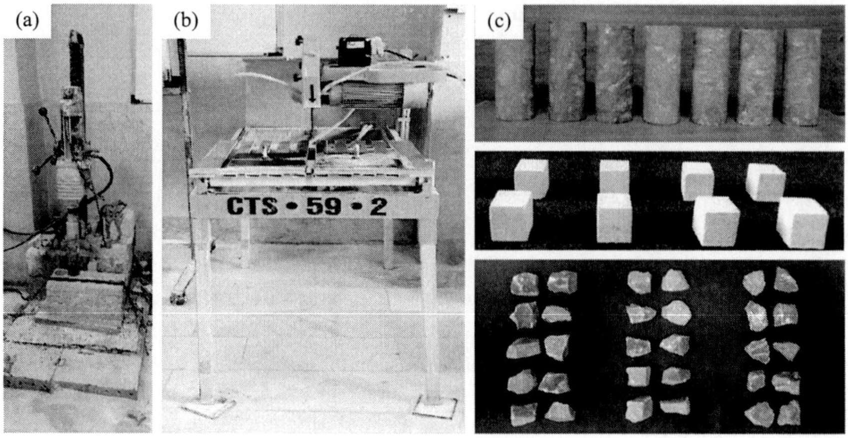

Figure 2.2 Images of (a) laboratory rock coring machine, (b) laboratory rock cutting machine and (c) cylindrical, cubic and lump-shaped laboratory specimens.

cubic and rectangular cubic specimens. Figure 2.2 shows images of laboratory rock coring and cutting machines and specimens prepared in different shapes. In the preparation of all laboratory specimens, they should be adjusted according to the standard suggested methods of tests.

References

ASTM, 1996. *Annual Book of ASTM Standards, Soil and Rock, Construction*. Vol. 8, Section 4. ASTM, West Conshohocken, PA, 950 pp.

ASTM, 2001. *Standard Practice for Preparing Rock Core Specimens and Determining Dimensional and Shape Tolerance. Designation:* 4543-01. ASTM, West Conshohocken, PA.

Fereidooni, D., 2023. *Engineering Geology*. Damghan University Press, Damghan, 552 pp.

Fereidooni, D., Askari, Q., 2018. Selection of the optimal direction of Spiro water transfer tunnel excavation based on orientation of rock mass discontinuities. *The First Biennial National Conference on the Geology of the Alborz Mountains and the Caspian Sea* November 29 and 30, 2018, Golestan University, Gorgan, Iran.

Fereidooni, D., Khajevand, R., (translators), 2019. *Rock Mechanics: An Introduction*. Damghan University Press, Damghan, 280 pp.

ISRM, 2007. Compilation arranged by the ISRM Turkish National Group, Ankara, Turkey. In: Ulusay, R., Hudson, J.A. (eds) *The Blue Book: The Complete ISRM Suggested Methods for Rock Characterization, Testing and Monitoring, 1974–2006*. Kazan Offset Press, Ankara.

Khodadad-Zadeh, M., Fereidooni, D., Diamantis, K., 2022. An engineering geological assessment for the Darband dam site, NE of Iran, using eight rock mass classification systems. *Innovative Infrastructure Solutions*, 2022, 7(2), 1–16.

Stölben, F., Eitner, V., 2004. Standardized methods for sampling by drilling and excavation and for groundwater measurements. In: Hack, R., Azzam, R., Charlier, R. (eds) *Engineering Geology for Infrastructure Planning in Europe.* Lecture Notes in Earth Sciences, vol 104 (pp. 101–109). Springer, Berlin, Heidelberg.

Chapter 3

Petrographic description and physical properties

3.1 Introduction

After transferring the obtained rock samples to the laboratory, the first actions that should be performed on them are the petrographic description, identifying mineralogical characteristics, determining texture, structure and lithology, as well as evaluating their physical characteristics. Although a lot of information may have been obtained about these characteristics in field studies, they should be examined more precisely in the laboratory. The mineralogical and lithological properties of rocks can influence many engineering properties, especially their physical properties, because, for example, the minerals that make up a rock, its texture and the presence of micro-cracks and fissures in the rock structure can affect its density and porosity. For this reason, knowledge of the mineralogical, lithologic and physical characteristics of rock is considered the most important preliminary stage of studies. These characteristics are directly related to other studies of rocks such as investigation of strength and deformation, durability, wearability, hardness, swelling, classification and operational studies such as drilling, mining, preparation of stony building materials, stability of rock structures (slopes and tunnels) and rock mass strength and classification.

3.2 Rock description in the hand specimen

Different rocks in the hand specimen have several characteristics that should be taken into consideration when examining them from an engineering point of view. The theoretical description of the rocks in the hand specimen is performed based on the suggested methods of ASTM (2009, 2022) and ISRM (2007) standards. When describing a rock in the hand specimen, all visible features such as appearance color; the fresh surface color of the rock; density; composing materials; size, arrangement and placement of grains and crystals; roundness or angularity of the grains; hardness; and the presence of joints, fractures, cavity and possible veins on the rock surface should be carefully investigated. By breaking the rock using a geological hammer, the new surface can be accessed and the mentioned parameters can be checked on the new surface. In this regard, a geological loop can be used to identify the type, shape and arrangement of grains and rock crystals. Also, to determine the qualitative hardness of the rock and its constituent minerals, the hardness of the constituent minerals can be determined by using a piece of flat glass and a knife blade using the Mohs hardness scale. By dragging a rock on an unglazed porcelain, its streak color can be determined to identify the minerals. The use of dilute hydrochloric acid (0.1 molar) helps to identify carbonate minerals such as calcite and dolomite. After describing the hand specimen, based on the examined items, the rock type and its name should be stated.

DOI: 10.1201/9781003672920-3

Figure 3.1 Characteristics of two rock types in the hand specimens: (a) hornfels and (b) limestone.

For example, according to Figure 3.1, the description of the two cylindrical and lumpy hand specimens is as follows.

Specimen A has a gray appearance and a dark gray color on its fresh surface. This specimen has clear and parallel dark and light laminations, which are caused by its parent rock. There are no cracks or fractures in it. It has a microcrystalline texture without surface cavities. The density of the rock in the hand specimen is high. This specimen is a metamorphic rock called hornfels.

Specimen B has a brown appearance and a fleshy color on its fresh surface. This specimen has a clear layering with a thickness of about 3 cm, and random and irregular joints and fractures can be seen in it, which in some cases are filled with fine-grained soils and iron oxides. It has a crystalline texture, and its grains are medium in size. This specimen has no surface cavities. This rock is without the surface cavity. The rock density in the hand specimen is relatively high. This specimen is a sedimentary rock called limestone.

3.3 Rock examination in the microscopic thin section

Rock examination in microscopic thin sections is done based on the suggested methods of ASTM (2009, 2022) and ISRM (2007) standards. To examine the rock on a microscopic scale, first, a microscopic thin section should be prepared from it. Then, the prepared thin section is examined using a polarizing microscope in polarized and normal lights, and its constituent minerals and texture should be determined. In the next step, a microscopic colored image or images can be taken from the thin section with suitable magnification, and the condition of the rock-composing minerals can be determined on each photo. It is suggested to prepare thin sections in several different directions for anisotropic and laminated rocks. After studying the rock under a microscope, the rock type and its name should be determined based on the classifications provided for the igneous, sedimentary and metamorphic rocks. It should be noted that in the study of texture, characteristics such as the size of grains or crystals composing the

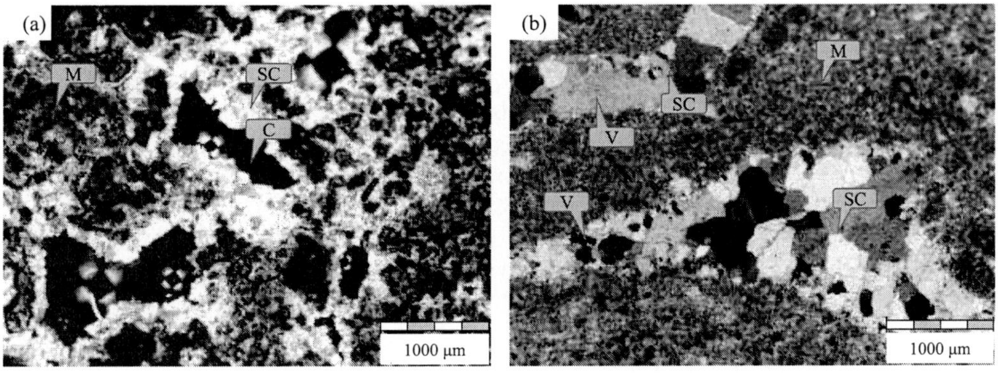

Figure 3.2 Characteristics of two rock types in microscopic thin sections: (a) travertine and (b) limestone (SC, sparry calcite; M, micrite; C, cavity; V, vein; F, fossil).

rock, their arrangement, contact of the grains with each other and the state of cavities and possible fractures in the rock specimen are all determined (Fereidooni and Sousa 2022). Figure 3.2 shows the characteristics of two rock types in microscopic thin sections.

3.4 Physical properties of rocks

Physical properties are the response of materials against intrinsic factors and the surrounding physical environments, which are measured in different fields and disciplines for these materials. For example, in rock mechanics, engineering geology, construction materials and geophysics, various physical properties are measured for rocks. The most important physical properties calculated in rock mechanics include density and unit weight in dry and saturated states, water content, specific gravity, porosity, void ratio, water absorption and saturation degree (Fereidooni 2016). Physical properties can be classified into two groups. The first group includes internal or intrinsic physical properties that are affected by the rock itself. The second group of physical properties are influenced by the surrounding environment. From the first group, we can refer to the porosity and void ratio, and from the second group, we can mention the unit weight and water absorption. In the following, the most important physical properties of rocks will be described in detail.

3.4.1 Density

The density of any object is the mass of its unit volume, which is represented by ρ and its unit is kg/m^3 or g/cm^3. The density of rocks depends on mineral composition and their constituent elements as well as the compaction degree during rock formation and its time. In general, igneous and metamorphic rocks are denser than sedimentary rocks. The density of rocks can be calculated using the following equation:

$$\rho = \frac{m}{V} \tag{3.1}$$

where m is the mass of the rock specimen and V is its volume. Density can be measured for rocks in five states, including solid particle density, dry density, moist density, saturated density and submerged density. For this purpose, it is enough to divide the dry mass, moist mass and saturated mass of the rock by its volume to calculate the mentioned parameters. A suitable method for statistical evaluation of rock density is to draw a histogram of the data set's statistical range. The numerical value of density and its changes for the data set can be (i) mean (average value), (ii) mode (the most common value or the peak of the distribution curve), (iii) median (the middle sample value of the data set, i.e. the value that is half of the samples are below and half above it) and (iv) the standard deviation (a statistical measure for the distribution of the data).

3.4.2 *Unit weight*

Sometimes, instead of density, another parameter named unit weight is used, which is represented by γ and its unit is N/m^3. The unit weight can be calculated from the following equation:

$$\gamma = \frac{W}{V} \tag{3.2}$$

where W is the weight of the rock specimen. Like density, the unit weight is also measured for rock in three states, including dry, moist and saturated. The unit weight and density are related to each other according to the following relationship:

$$\gamma = \rho \times g \tag{3.3}$$

where g is the earth's gravitational acceleration, which is equal to 9.81 m/s^2. The unit weight is a function of the amount of rock voids, mineralogical composition, depth of formation and geological age of the rock, and it may be measured as solid part, dry, moist, saturated and submerged unit weights; their measurement equations are presented in Table 3.1.

Table 3.1 Different states of unit weights

No.	*Parameter*	*Equation*
1	Solid part unit weight	$\gamma_s = \frac{W_d}{V_s}$
2	Dry unit weight	$\gamma_d = \frac{W_d}{V_T}$
3	Moist unit weight	$\gamma_m = \frac{W_m}{V_T}$
4	Saturated unit weight	$\gamma_{sat} = \frac{W_{sat}}{V_T}$
5	Submerged unit weight	$\gamma_{sub} = \gamma' = \gamma_{sat} - \gamma_w$

In this table, W_d is rock weight in dry state, which is due to the weight of solid particles; W_m is rock weight in the moist state; W_{sat} is rock weight in the saturated state; and V_T is total volume of the rock, which is equal to volume sum of the solid particles, V_s, and the empty voids V_v. γ_w is the water unit weight.

Among the most important factors affecting rock unit weight, the following can be mentioned:

- Mineral types composing rock: In this regard, the presence of dense minerals in rock composition increases its unit weight.
- Compaction degree of rock: With an increase in rock compaction degree, the amount of its unit weight increases.
- Empty voids and small fractures in rock (porosity): As the volume of voids in rock increases, its dry unit weight decreases.
- Factors affecting rock porosity: All factors that increase rock porosity reduce its unit weight.
- Rock cementation degree and recrystallization: Increasing the degree of cementation and recrystallization of rock increases its unit weight.

3.4.3 *Specific gravity*

The specific gravity is the ratio of the weight of a unit volume of any object to the weight of the unit volume of pure water at 4°C, which is a dimensionless quantity. Since a unit volume of pure water at 4°C is equal to 1 g/cm^3, the specific gravity is equal to rock unit weight. If for any reason the unit weight of water changes from the mentioned value, the rock unit weight and specific gravity are not equal to each other. The specific gravity is considered an important feature in the construction of engineering projects because, for example, in the construction of anchorages and coastal walls, it is important to have walls with a high specific gravity. After all, the stones used in the construction of these structures are immersed in water and they lose an important value of their effective weight. On the other hand, in the construction of some structures such as arches or domes, it is necessary to use light stones. The specific weight can be calculated by the following equation:

$$G = \frac{\gamma}{\gamma_w} \tag{3.4}$$

In this equation, γ is the unit weight of rock and γ_w is the unit weight of water.

3.4.4 *Solid-particle specific gravity*

The solid-particle specific gravity is the ratio of the solid particle unit weight of rock to the unit weight of pure water at 4°C, which is a dimensionless quantity. The solid particle specific gravity can be calculated using the following equation:

$$G_s = \frac{\gamma_s}{\gamma_w} \tag{3.5}$$

In this equation, γ_s is the unit weight of solid rock particles and γ_w is the water unit weight.

3.4.5 *Porosity*

Rocks always contain voids in the form of connected or separate cavities and cracks. According to the definition, porosity is the ratio of the volume of empty spaces in a rock to the total rock volume, which is expressed as a percentage and can be calculated using the following equation:

$$n = \frac{V_v}{V} \times 100 \tag{3.6}$$

In this equation, n is the porosity, V_v is the volume of empty voids and V_T is the total volume of rock. The porosity can be classified from different points of view. From a geological point of view and based on the formation time of voids in rock, it is classified into two types including primary and secondary porosity. The primary porosity is created during the formation of rock, such as empty voids between the grains of a sedimentary rock or cavities in a porous volcanic rock such as pumice. The primary porosity includes intercrystalline porosity, intergranular porosity, interlayer porosity and organic porosity. The secondary porosity occurs after the formation of rocks and as a result of subsequent geological processes, such as the creation of joints, cracks and faults, dissolution and the dolomitization process in rocks.

From an engineering point of view, total porosity includes effective and ineffective porosities. The effective porosity consists of interconnected voids that are capable of allowing fluids to pass through them. This porosity is obtained from the ratio of the volume of the interconnected holes to the total volume of rock. It should be noted that there are holes in the rock that do not allow liquids to pass. These holes are called ineffective porosity, and it is obtained from the difference between effective porosity and total porosity.

From the point of view of void size, the porosity is divided into two types, including macroscopic and microscopic. In the macroscopic porosity, the pore diameter is more than 8 microns. In these pores, water and fluid materials flow constantly and easily, and their movement is subject to the force of gravity. The pores in this type of porosity are caused by the dissolution of materials, tectonic pressures or fractures caused by the drying of rocks and sediments. This type of porosity is often visible in alluvial sediments and rocks with high solubility such as limestone. In the microscopic porosity, the diameter of the pores is smaller than 8 microns. In these pores, water or other fluids cannot move easily. Therefore, in this type of pores, the force of gravity is not effective in the passage and flow of liquids. The movement of liquids in these pores is subject to the capillary tube law. The microscopic porosity is abundant in clay-bearing rocks. These rocks are impermeable and hold water in their small pores and lose it with difficulty. In fact, the porosity in the clay-bearing rocks is ineffective and the pores and holes in them are not connected.

The most important factors affecting the porosity include the following:

- Size of rock grains or crystals: Usually, the smaller the grain size, the greater the porosity.
- Sorting of rock grains: The more the grains are in the same size and the better their sorting, the greater the porosity and permeability.
- The shape of rock grains: Grains with more roundness and sphericity increase the porosity, while non-spherical and oriented particles decrease the porosity.
- Arrangement and orientation of rock grains: The arrangement of grains is effective in the degree of porosity and permeability. Spherical grains have two types of arrangement, namely, cubic and rhombohedral. If the grains have a cubic arrangement, the value of the void ratio is 0.91, and if they are rhombohedral, it is about 0.35.
- Compaction degree of rock: With the increasing compaction degree of rock, its porosity decreases.
- Cementation degree of rock: Increasing the cementation degree of rock during its formation reduces its porosity.

In Tables 3.2 and 3.3, a classification of intact rock based on the dry unit weight and porosity and the value ranges of dry unit weight and porosity of some rocks are presented.

Table 3.2 Classification of intact rock based on dry unit weight and porosity (Anon 1979)

Class	*Dry unit weight (t/m³)*	*Description*	*Porosity (%)*	*Description*
1	<1.8	Very low	>30	Very high
2	1.8–2.2	Low	15–30	High
3	2.2–2.55	Moderate	5–15	Medium
4	2.55–2.75	High	1–5	Low
5	>2.75	Very high	<1	Very low

Table 3.3 Value ranges of physical and index properties of some rocks

Group	*Rock type*	*Dry unit weight (g/cm³)*	*Porosity (%)*
Igneous	Granite	2.53–2.62	1.02–2.87
	Diorite	2.80–3.00	0.10–0.50
	Gabbro	2.72–3.00	1.00–3.57
	Rhyolite	2.40–2.60	0.40–4.00
	Andesite	2.50–2.80	0.20–8.00
	Basalt	2.21–2.77	0.22–22.10
Sedimentary	Conglomerate	2.47–2.76	–
	Sandstone	1.91–2.58	1.62–26.40
	Shale	2.00–2.40	2.00–50.00
	Mudstone	1.82–2.72	–
	Dolomite	2.20–2.70	0.20–4.00
	Limestone	2.67–2.72	0.27–4.10
Metamorphic	Gneiss	2.61–3.12	0.32–1.16
	Schist	2.60–2.85	10.00–30.00
	Phyllite	2.18–3.30	–
	Slate	2.71–2.78	1.84–3.64
	Marble	2.51–2.86	0.65–0.81
	Quartzite	2.61–2.67	0.40–0.65

3.4.6 *Void ratio*

The void ratio is the ratio of the voids' volume in a rock specimen to the volume of its solid particles, which is expressed as a decimal number and can be calculated through the following equation:

$$e = \frac{V_v}{V_s} \tag{3.7}$$

In this equation, e is the void ratio, V_v is the volume of voids in the rock and V_s is the volume of solid materials in the rock. The following relationships exist between the porosity and the void ratio:

$$e = \frac{n}{1-n} \tag{3.8}$$

$$n = \frac{e}{1+e} \tag{3.9}$$

3.4.7 *Water content*

The water content or moisture percentage is a ratio of water amount in a rock specimen that may have entered it naturally or unnaturally to the dry weight of the specimen, which can be calculated from the following equation:

$$\omega = \frac{W_w}{W_s} \times 100 \tag{3.10}$$

In this equation, W_w is the water weight in the specimen and W_s is the weight of the solid particles of the specimen.

3.4.8 *Water absorption*

Water absorption expresses the maximum amount of water that a rock specimen can absorb to reach the saturation degree. Therefore, the volume of absorbed water in this case is equivalent to the volume of voids, or the porosity of the rock. In other words, water absorption is the percentage of rock moisture when the rock is saturated with water. The amount of water absorption in rocks is very different, and sometimes there may be many changes in the water absorption in one rock type. But in general, the water absorption of igneous rocks is much lower than that of metamorphic and sedimentary rocks, respectively. The rock water absorption can be calculated from the following equation:

$$W_a = \frac{W_w}{W_s} \times 100 \tag{3.11}$$

In this equation, W_w is the weight of water in the rock specimen and W_s is the weight of the solid particles of the specimen.

3.4.9 *Saturation ratio*

The saturation ratio is the ratio of the volume of water in the voids of a rock specimen to the volume of the voids, which is expressed as a percentage. In other words, the saturation ratio indicates how much water is filled in the rock cavities. The numerical value of this parameter varies between zero and 100%. The saturation ratio is calculated from the following equation:

$$S_r = \frac{V_w}{V_v} \times 100 \tag{3.12}$$

In this equation, V_w is the volume of water in specimen voids and V_v is the volume of the voids.

3.5 Determining the physical properties of rock in the laboratory

Determining the physical properties (weight–volume indices) of rock in the laboratory is done based on the methods proposed by ASTM (2010) and ISRM (1979, 2007). In this regard, two methods including measuring specimen dimensions and immersion in water are usually used. The first method is used for specimens that have a regular geometric shape, such as cylindrical, disc and cubic specimens, while the second method is used for specimens that do not have a regular geometric shape and whose dimensions cannot be measured.

3.5.1 Method of measuring specimen dimensions

In the method of determining the physical properties of rock by measuring the specimen dimensions, first, the exact dimensions of the specimen are measured several times from its different parts with a caliper, and the obtained values are averaged. Then the weight of the specimen is measured in two states: dry and saturated. To measure the dry weight, the specimen must first be dried in an oven at a temperature of 105°C for 24 hours. Then, the specimen is weighed with a digital scale with an accuracy of 0.01 g. There are two methods for saturating the specimen. In the first method, the rock specimen is immersed in pure water at a temperature of 4°C at least for a week until it becomes saturated (water at a temperature of 4°C or 39.2°F has a density of 1 g/cm^3, 1 kg/m^3 and 62.4 lb/ft^3). Another method is to use a desiccator, by which the specimen can be saturated in 1 hour by placing it within the desiccator, which contains pure water at a temperature of 4°C, by creating a vacuum. Then, the specimen is removed from the water, its surface is dried with a moist cloth, and it is quickly weighed by a digital scale with an accuracy of 0.01 g (Shariati and Fereidooni 2021). For each rock type, five specimens should be tested to determine its weight–volume indices and the average obtained values should be calculated. Having the total volume of the rock specimen (V) and its dry and saturated weights (W_d and W_{sat}), the volumes of the solid part (V_s) and voids (V_v) in the rock specimen can be calculated as follows:

$$W_w = W_{sat} - W_d;\ \gamma_w = 1\frac{g}{cm^3} \rightarrow W_w = V_w = V_v \rightarrow \qquad (3.13)$$
$$V_v = W_{sat} - W_d$$

$$V_s = V - V_v \qquad (3.14)$$

By evaluating the above parameters, all the physical properties (weight–volume indices) of the rock can be calculated using the existing equations.

3.5.2 Specimen immersion method

In the method of determining the physical properties of the rock by the immersion method, the weight of rock specimen is measured in three states including dry and saturated in air and saturated in water. Measuring the dry and saturation weights of the specimen in air is done as mentioned before. The Archimedes scale is used to determine the saturated weight of the specimen while immersed in water. In all cases, the weight of the specimen is obtained with an accuracy of 0.01 g (Fereidooni and Khajevand 2023). For each rock type, five specimens should be tested to determine its weight–volume indices, and the average of the obtained values should be calculated. By determining the weight of the specimen in three states, namely, dry (W_d), saturated in air ($W_{sat(a)}$) and saturated in water ($W_{sat(w)}$), the total volume (V) of the specimen can be first calculated based on Archimedes' principle, from the following equation:

$$V = W_{sat(a)} - W_{sat(w)} \qquad (3.15)$$

Then, the volume of the solid parts (V_s) and the volume of the voids (V_v) in the rock specimen can be calculated from the following relations, similar to the previous state:

$$V_v = W_{sat(a)} - W_d \qquad (3.16)$$

$$V_s = V - V_v \qquad (3.17)$$

In this case, by evaluating the above parameters, all the physical properties (weight–volume indices) of the rock can be calculated using the existing relationships. The most important equations between physical properties (weight–volume indices) of rock are presented in Table 3.4.

While measuring the physical properties of rock in the laboratory, it should be noted that many factors influence the accuracy of the results. These factors include intrinsic and internal factors of the rock, laboratory factors and environmental factors that should be considered.

Table 3.4 The most important equations between physical properties (weight–volume indices) of rock

Parameter	*Equation*
Dry unit weight	$\gamma_d = \frac{\gamma}{1+\omega}$
	$\gamma_d = \frac{G_s \gamma_w}{1+e}$
	$\gamma_d = G_s \gamma_w (1-n)$
	$\gamma_d = \frac{G_s}{1+\frac{\omega G_s}{S}} \cdot \gamma_w$
	$\gamma_d = \frac{e S \gamma_w}{(1+e)\omega}$
	$\gamma_d = \gamma_{sat} - n\gamma_w$
	$\gamma_d = \gamma_{sat} - \left(\frac{e}{1+e}\right)\gamma_w$
Moist unit weight	$\gamma_m = \frac{G_s \gamma_w (1+\omega)}{1+e}$
	$\gamma_m = \frac{(G_s + Se)\gamma_w}{1+e}$
	$\gamma_m = \frac{G_s \gamma_w (1+\omega)}{1+\frac{\omega G_s}{S}}$
	$\gamma_m = G_s \gamma_w (1-n)(1+\omega)$

(*Continued*)

Table 3.4 (Continued)

Parameter	*Equation*
Saturated unit weight	$\gamma_{sat} = \frac{(G_s + e)\gamma_w}{1+e}$
	$\gamma_{sat} = [G_s(1-n)+n]\gamma_w$
	$\gamma_{sat} = \left(\frac{1+\omega}{1+\omega G_s}\right)G_s\gamma_w$
	$\gamma_{sat} = \left(\frac{e}{\omega}\right)\left(\frac{1+\omega}{1+e}\right)\gamma_w$
	$\gamma_{sat} = \gamma_d + n\gamma_w$
	$\gamma_{sat} = \gamma_d + \left(\frac{e}{1+e}\right)\gamma_w$

References

Anon, 1979. Classification of rocks and soils for engineering geological mapping, Part 1: Rock and soil materials. Report of the Commission of Engineering Geological Mapping. *Bulletin of the International Association of Engineering Geology*, 19, 364–371.

ASTM, 2009. *Standard Guide for Petrographic Examination of Dimension Stone (C1721). Book Standards, Vol. 04.07.* ASTM, West Conshohocken, PA.

ASTM, 2010. *Designation D2216-10: Standard Test Methods for Laboratory Determination of Water (Moisture) Content of Soil and Rock by Mass*. ASTM International, West Conshohocken, PA.

ASTM-C1721, 2022. *Standard Guide for Petrographic Examination of Dimension Stone*. ASTM International, West Conshohocken, PA.

Fereidooni, D., 2016. Determination of the geotechnical characteristics of hornfelsic rocks with a particular emphasis on the correlation between physical and mechanical properties. *Rock Mechanics and Rock Engineering*, 49(7), 2595–2608.

Fereidooni, D., Khajevand, R., 2023. Evaluation of engineering properties of travertines and their relationships – A case study: Babahafez quarry of Damghan. In *The 25th Conference of Geological Society of Iran* 25th and 26th January 2023, Shahrood University of Technology, Shahrood, Iran.

Fereidooni, D., Sousa, L., 2022. Predicting the engineering properties of rocks from textural characteristics using some soft computing approaches. *Materials*, 15, 7922.

ISRM, 1979. Suggested method for determining water content. Porosity, density, absorption and related properties and swelling and slake-durability index properties. *International Journal of Rock Mechanics and Mining Sciences & Geomechanics Abstracts*, 16(2), 141–156.

ISRM, 2007. Compilation arranged by the ISRM Turkish National Group, Ankara, Turkey. In: Ulusay, R., Hudson, J.A. (eds) *The Blue Book: The Complete ISRM Suggested Methods for Rock Characterization, Testing and Monitoring, 1974–2006*. Kazan Offset Press, Ankara.

Shariati, M., Fereidooni, D., 2021. Rock slope stability evaluation using kinematic and kinetic methods along the Kamyaran-Marivan road, West of Iran. *Journal of Mountain Science*, 18(3), 779–93.

Chapter 4

Wave velocity and dynamic elastic constants

4.1 Introduction

The ultrasonic wave velocity test in rocks is a fast, non-destructive and inexpensive method that can be used both in the laboratory and on the site. Direct tests to measure the strength and engineering properties of rocks are generally expensive and destructive and require a long time for specimen preparation and testing. Therefore, measuring the velocity of ultrasonic waves is a suitable method for determining and estimating different characteristics of rocks, especially when cylindrical specimens are required for direct tests. The ultrasonic wave velocities are often used to determine the dynamic elastic constants of rocks. Also, by using the velocity of ultrasonic waves in the rocks, it is possible to determine the quality index (IQ), evaluate the porosity and fissures of the rocks and estimate some properties such as the uniaxial compressive strength, tensile strength, shear strength, impact value index and Schmidt rebound hardness using experimental relations (Khajevand and Fereidooni 2018). In general, it is possible to evaluate some properties of rocks using the laboratory test results of the ultrasonic wave velocities as follows:

- Evaluation of the uniformity and homogeneity of the rock
- Identifying the location of cracks, fractures and cavities within the rock
- Evaluation of the inherent and induced anisotropy of the rock
- Investigating the water presence in the rock
- Determining the dynamic elastic constants of the rock

The ultrasonic waves are classified into two types: primary wave (P-wave) or longitudinal compression wave, and secondary wave (S-wave) or transverse shear wave. The P-wave, while passing through the material body (solid and liquid), causes its volume to change, and its propagation direction coincides with its vibration direction. The propagation velocity of the P-wave is higher than the S-wave. The velocity of this wave in rocks is about several kilometers per second, which depends on the elastic properties of the rock. The S-wave passes through the rock and causes its shear deformation. The propagation direction of this wave is perpendicular to its vibration, and the wave velocity is lower than the P-wave (Fereidooni and Khajevand 2016). Since liquids do not withstand shear stress, the S-wave does not pass through the liquids.

4.2 Determining ultrasonic wave velocities in rocks

In the laboratory, the measurement of the ultrasonic wave velocities in rocks is performed based on the methods proposed by ASTM (1996) and ISRM (2007) standards. Since this test is a

DOI: 10.1201/9781003672920-4

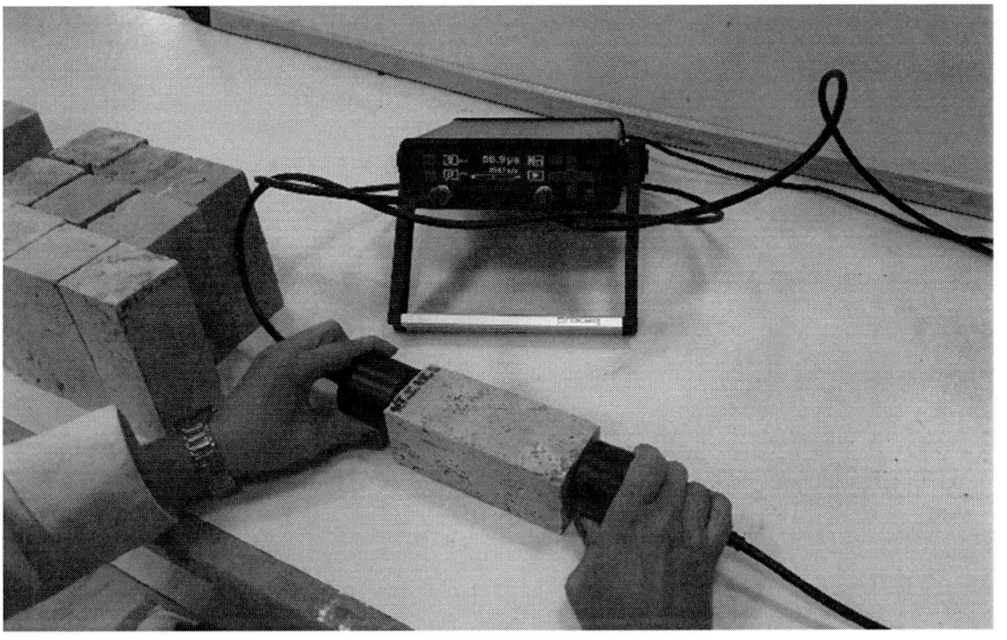

Figure 4.1 Measuring the velocity of P- and S-ultrasonic waves with the Pandit device.

nondestructive method, it has been widely used in geological and geotechnical engineering. Figure 4.1 shows the device and test method to determine the wave velocity in rocks. The method of using wave velocity in rocks is the most suitable method for determining their dynamic elastic constants (ASTM 1996). The laboratory device for measuring the ultrasonic wave velocities in rocks is the Pandit device, which is used to measure the velocity of both P- and S-wave velocities in rocks, concrete, ceramic and wood samples. To use this device, it must firstly be calibrated. The device has both P- and S-wave generators. The P-wave generator measures the P-wave velocity passing through the rock specimen and has a frequency of 54 kHz, which, when converting electrical energy to ultrasonic energy, sends waves with a shorter wavelength and penetrates to a lower depth. The S-wave generator measures the S-wave velocity passing through the rock specimen and has a frequency of 500 kHz. According to the frequency of this converter, it can be said that this type of converter sends longer waves that penetrate deeper when converting electrical energy to ultrasonic energy. High-frequency converters are suitable for small, loose and non-homogeneous specimens, while low-frequency converters are suitable for large, dense and relatively homogeneous ones.

When measuring the P- and S-ultrasonic wave velocities, the arrangement of transducers can be in the following three ways (Figure 4.2):

- Direct arrangement: In this mode, there is the highest amount of signal transmission between the transducers.
- Semi-direct arrangement: It is less sensitive than the previous one, but more sensitive than the indirect one.
- Indirect arrangement: This method is used to determine crack depth and surface quality.

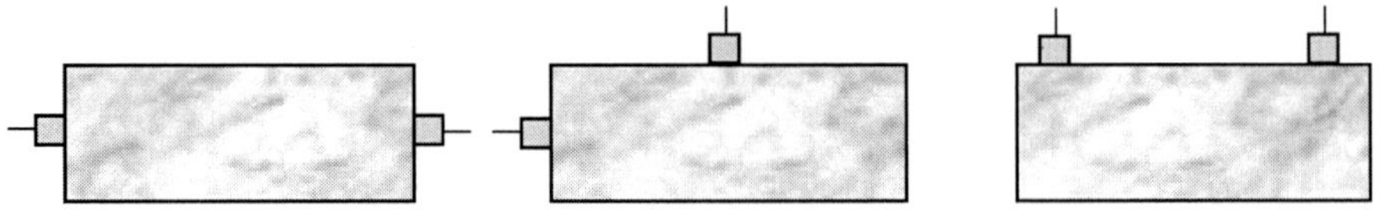

Figure 4.2 Arrangement of transducers in the Pandit device.

Table 4.1 Classification of rocks based on the ultrasonic wave velocity (Anon 1979)

Class	*P-wave velocity (m/s)*	*Description*
1	<2500	Very low
2	2500–3500	Low
3	3500–4000	Medium
4	4000–5000	High
5	>5000	Very high

For determining the ultrasonic wave velocity, the transducers are completely connected to the surface of the rock specimen at the entering and exiting points of the wave. The transducers' places on the rock specimen can have a different arrangement according to the test purpose. The transducers are completely connected on the smooth and polished surface of the rock using silicone gel to prevent energy loss and ensure energy transfer to the specimen. By determining the test time, the wave passes through the specimen (the time from the arrival to the exit of the wave, t), which is measured by the device, and the length of the specimen (L), which is measured by a caliper, and the ultrasonic wave velocity in the rock specimen is calculated from the following equation:

$$v = \frac{L}{t} \tag{4.1}$$

To correctly perform the test to determine the ultrasonic wave velocities, it is necessary to observe the following points:

- Before conducting the test, the length of the rock specimen should be accurately measured from several parts, and its average value should be calculated.
- Before performing the test, the device should be calibrated and set to zero.
- During the test, the specimen should be placed in a fixed place and no pressure should be applied to it because it causes an error in the results.
- The transducers are placed in their locations, and the number displayed on the device is read and recorded.
- The velocity measurement should be done in a specific direction for all specimens.

Based on the ultrasonic wave velocities, different rocks are classified according to Table 4.1. In Table 4.2, the range of P- and S-wave velocities of rocks is presented for comparison.

Table 4.2 Range of P- and S-wave velocity values of some rocks

Rock group	*Rock type*	*P-wave velocity (m/s)*	*S-wave velocity (m/s)*
Igneous	Granite	4500–6000	3500–3800
	Diorite	4500–6700	–
	Gabbro	4500–7000	–
	Andesite	4500–6500	–
	Basalt	5000–7000	3660–3700
Sedimentary	Conglomerate	–	3500–3700
	Sandstone	1500–4600	–
	Shale	200–4600	–
	Dolomite	5500	–
	Limestone	3500–6500	–
Metamorphic	Gneiss	5000–7500	–
	Schist	6100–6700	3460–4000
	Slate	3500–4500	–
	Marble	5000–6000	–

4.3 Factors affecting the ultrasonic wave velocities in rocks

The most important factors affecting the ultrasonic wave velocities in rocks are as follows:

- Rock type: In rocks with a higher degree of density, which often include igneous and metamorphic rocks, the wave velocity is higher.
- Rock texture: The wave velocity is affected by the size of grains composing the rock, so that the wave velocity of fine-grained rocks is generally higher than that of coarse-grained rocks.
- Density: Generally, the increase in density increases the wave velocity, and the relationship between the two parameters may be linear or non-linear.
- Porosity: The wave velocity decreases with increasing porosity, and this phenomenon is true in both dry and saturated rocks. Experience has shown that the shear wave velocity decreases rapidly with increasing porosity.
- Anisotropy: The angle between the anisotropy surfaces and the direction of wave propagation is one of the important factors affecting the wave velocity. The wave velocity along the parallel direction to the anisotropy surfaces in both saturation and natural moisture conditions is always higher than the wave velocity in the perpendicular direction to these surfaces.
- Stress level: As the stress level increases, the wave velocity also increases. At low stresses, a sharp increase in wave velocity is associated with the reduction of voids, the blocking of cracks and fine fractures and the increase in the mechanical contact of constituent grains. The increase in the wave velocity at higher stresses is the result of changes in some of the inherent characteristics of rocks, such as the creation of cracks in the crystal network.
- Moisture content: The rock moisture usually increases the P-wave velocity, but since the S-wave passes only through the solid part of the rock, the moisture content does not have much effect on its velocity. If the rock is assumed to be completely saturated, the velocity of waves in rocks with high porosity is lower than in rocks with low porosity, because the velocity of P-waves in water is lower than in solid rocks.

- Type and state of pore fluid: The presence of pore fluids with a higher density within the rock porosity increases the wave velocity in it. Also, the freezing of water as a pore fluid inside the rock cavities increases the wave velocity in it (Arab-Ameri and Fereidooni 2023).
- Temperature: Usually, with increasing rock temperature, the wave velocity in the rock and its elastic constants decrease.
- Degree of rock weathering: As the degree of rock weathering increases, the wave velocity decreases.

4.4 Dynamic elastic constants of rocks

The dynamic elastic constants of rock include modulus of elasticity or Young's modulus (*E*), shear or rigidity modulus (*G*), Poisson's ratio (υ), bulk or incompressibility modulus (*K*) and Lame's constant (λ). The elastic constants can be measured statically or dynamically for rocks. It should be mentioned that the static elastic constants are determined by conducting strength/deformational laboratory tests, while dynamic elastic constants can be determined indirectly using the P- and S-wave velocities in rock. By determining the P- and S- ultrasonic wave velocities for a rock sample, its dynamic elastic constants can be obtained through the following relations (Del Rio et al. 2007; ASTM 2000):

$$E = \frac{\rho.v_s^2\left(3v_p^2 - 4v_s^2\right)}{v_p^2 - v_s^2} \tag{4.2}$$

$$G = \rho.v_s^2 \tag{4.3}$$

$$v = \frac{v_p^2 - 2v_s^2}{2(v_p^2 - v_s^2)} \tag{4.4}$$

$$K = \frac{\rho\left(3v_p^2 - 4v_s^2\right)}{3} \tag{4.5}$$

$$\lambda = \rho\left(v_p^2 - 2v_s^2\right) \tag{4.6}$$

In these equations, v_p and v_s are the P- and S-wave velocities of the rock sample, respectively, and ρ is its density. The following relationships are established between the dynamic elastic constants of rock (Beer et al. 2012):

$$G = \frac{E}{2\left(1+v\right)} \tag{4.7}$$

$$K = \frac{E}{3\left(1-2v\right)} \tag{4.8}$$

$$\lambda = \frac{vE}{\left(1+v\right)\left(1-2v\right)} \tag{4.9}$$

On the other hand, knowing the dynamic elastic constants of a rock, its P- and S-wave velocities (v_p and v_s) can be calculated from the following relations (quoted from Fereidooni 2023):

$$v_p = \sqrt{\frac{\lambda + 2G}{\rho}} \tag{4.10}$$

$$v_s = \sqrt{\frac{G}{\rho}} \tag{4.11}$$

In the case that Poisson's ratio is assumed to be 0.25 for a rock, the following relations exist between the P- and S-wave velocities (quoted from Fereidooni 2014):

$$v_p = \sqrt{3} v_s \tag{4.12}$$

$$v_{p(\mathrm{m/s})} = 1.4 v_s + 657 \tag{4.13}$$

$$v_{s(\mathrm{m/s})} = 0.58 v_p + 210.61 \tag{4.14}$$

4.5 Determining rock quality using ultrasonic wave velocities

Fissures and micro-fractures in intact rock are the most important parameters affecting the ultrasonic wave velocities. The presence of fissures and micro-fractures increases the porosity of the rock and decreases the ultrasonic wave velocities in it. Therefore, the ultrasonic wave velocities can be related to the features in the intact rock. With the ultrasonic wave velocities, it is possible to estimate the rock quality. For this purpose, Fourmaintraux (1976) and Tourenq et al. (1971) have proposed two methods. According to Fourmaintraux (1976), the following equation can be used to calculate the IQ of rocks:

$$\mathrm{IQ} = \frac{v_p}{v_p^*} \times 100 \tag{4.15}$$

In this equation, v_p is the P-wave velocity measured from the tests of the rock sample and v_p^* is the P-wave velocity calculated based on the P-wave velocity of minerals composing the rock sample, which is obtained from the following equation:

$$v_p^* = \frac{1}{\sum \left(\frac{C_i}{v_{pi}}\right)} \tag{4.16}$$

In this equation, C_i and v_{pi} are the percentage of rock-forming minerals and the P-wave velocity of each mineral, respectively. The v_{pi} values for common rock-forming minerals are presented in Table 4.3.

Based on laboratory measurements and microscopic observations of fissures and micro-fractures in a rock specimen, Fourmaintraux (1976) proposed a plot of the rock IQ versus its porosity (n) as a basis for describing the degree of fissuring of the rock (Figure 4.3). This diagram has

Table 4.3 P-wave velocity values for common rock-forming minerals (Goodman 1989; Fereidooni 2018)

No.	*Mineral*	*P-wave velocity (m/s)*
1	Quartz	6050
2	Orthoclase	5800
3	Plagioclase	6250
4	Amphibole	7200
5	Olivine	8400
6	Biotite	5900
7	Muscovite	5800
8	Garnet	6500
9	Andalusite	5750
10	Sillimanite	5200
11	Staurolite	5700
12	Kyanite	5600
13	Graphite	4700
14	Pyrite	8000
15	Epidote	7450
16	Gypsum	5200
17	Magnetite	7400
18	Dolomite	7500
19	Calcite	6600

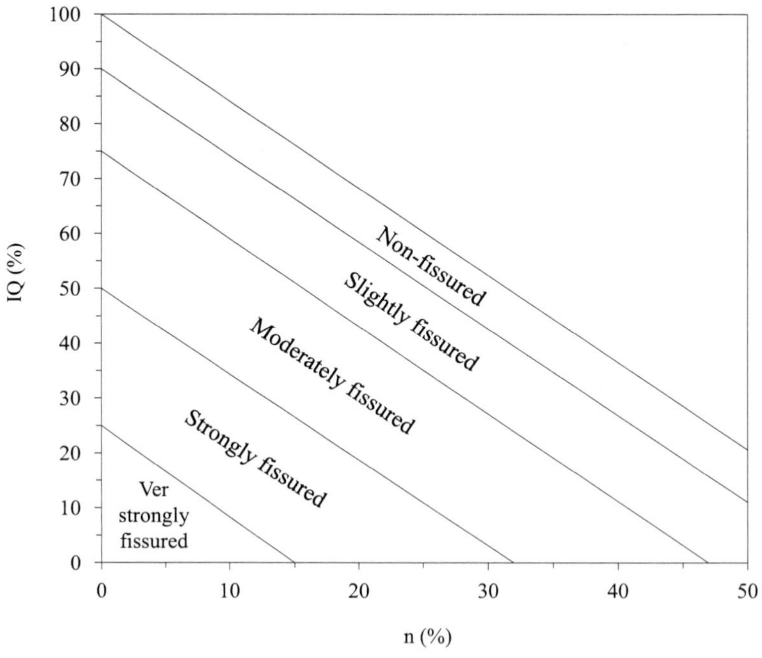

Figure 4.3 Intact rock classification diagram based on its fissuring degree (Fourmaintraux 1976).

five separate areas, including (i) non-fissured, (ii) slightly fissured, (iii) moderately fissured, (iv) strongly fissured and (v) very strongly fissured. Fourmaintraux (1976) stated that the relationship between the rock IQ and its porosity is as follows:

$$\text{IQ} = 100 - 1.6n\left(\%\right) \tag{4.17}$$

Fereidooni (2018) obtained the relationship between these two parameters as follows:

$$\text{IQ} = 100.36 - 12.56n\left(\%\right) \tag{4.18}$$

According to Tourenq et al. (1971), the rock fissuring degree can be determined from the ratio of S-wave velocity (v_s) to P-wave velocity (v_p). For rocks with strong fissures, this ratio is higher than 70%; for rocks with low fissures, it is between 60% and 70%; and for rocks without fissures, it is less than 60%.

References

Anon, 1979. Classification of rocks and soils for engineering geological mapping, Part 1: Rock and soil materials, Report of the Commission of Engineering Geological Mapping. *Bulletin of International Association of Engineering Geology*, 19, 364–371.

Arab-Ameri, S., Fereidooni, D., 2023. Assessing the parameters affecting on ultrasonic wave velocity of some building stones used in Iran. *Journal of Engineering Geology*, 17(1), 125–146.

ASTM, 1996. *Standard Test Method for Laboratory Determination of Pulse Velocities and Ultrasonic Elastic Constants of Rock. Designation: D2845-95*. ASTM International, West Conshohocken, PA.

ASTM, 2000. *Standard Test Method for Laboratory Determination of Pulse Velocities and Ultrasonic Elastic Constants of Rock, USA, Designation: D285-00*. ASTM International, West Conshohocken, PA.

Beer, F.P., Johnston, E.R., DeWolf, J., Mazurek, D.F., 2012. *Mechanics of Materials*. Sixth edition, McGraw-Hill Companies, Inc., New York, 838 pp.

Del Río, L.M., López, F., Calleja, B., Mota, M.I., Tejado, J.J., González, I., Emeterio, J.L., Ramos, A., 2007. Determination of dynamic elastic constants in granites using through–transmission A–scan and resonance–based ultrasonic techniques. *Proceedings of the International Congress on Ultrasonics*, Vienna, pp. 9–13.

Fereidooni, D., 2014. *General and Engineering Geology*. Daneshjoo Publications, Damghan, Iran, 450 pp. (in Persian).

Fereidooni, D., 2018. Assessing the effects of mineral content and porosity on ultrasonic wave velocity. *Geomechanics and Engineering*, 14(4), 399–406.

Fereidooni, D., 2023. *Engineering Geology*. Damghan University Press, Damghan, Iran, 552 pp. (in Persian).

Fereidooni, D., Khajevand, R., 2016. Determination of the correlation between physical properties of rocks using regression analysis. *Proceedings of the Second National Conference on Computational Sciences*, 7th and 8th September 2016, Damghan University, Damghan, Semnan, Iran, (in Persian).

Fourmaintraux, D., 1976. Characterization of rocks; laboratory tests: Chapter IV in La Mechanique des roches applique auxouvr ges du genie civil by Marc Panet et al., Ecole Nationale des Ponts et Chaussees.

Goodman, R.E., 1989. *Introduction to Rock Mechanics*. John Wiley & Sons Inc, New York, 562 pp.

ISRM, 2007. Compilation arranged by the ISRM Turkish National Group, Ankara, Turkey. In: Ulusay, R., Hudson, J.A. (eds) *The Blue Book: The Complete ISRM Suggested Methods for Rock Characterization, Testing and Monitoring, 1974–2006*. Kazan Offset Press, Ankara.

Khajevand, R., Fereidooni, D., 2018 Assessing the empirical correlations between engineering properties and P-wave velocity of some sedimentary rock samples from Damghan, northern Iran. *Arabian Journal of Geosciences*, 18, 528–547.

Tourenq, C., Fourmaintraux, D., Denis, A., 1971. Propagation des ondes etdiscontinuities des roches. *Proceedings of the International Society of Rock Mechanics, Symposium on Rock Fracture*. ISRM, Nancy, France.

Chapter 5

Durability and deterioration

5.1 Introduction

Weak rocks containing clay minerals such as mudstone, claystone, shale, marl and some volcanic rocks change their engineering properties due to changes in weather and climate factors, namely, temperature, humidity, wetting and drying, and show decay or deterioration over time. It is very important to have a suitable index to estimate the sensitivity of the rocks used in the projects built in areas with changing climates. Among the most important of such projects, there can be mentioned to piers, breakwaters, embankment dam shells, tunnel openings, bridges with stone foundations and masonry retaining walls. Also, stone structures and sculptures, stone facades of buildings and landscaping of roads and parks are always exposed to weather, temperature and air humidity changes, as well as the attack of acidic solutions (Fereidooni and Ghasemi 2025). In this regard, various methods have been presented to determine the durability and deterioration of rocks against such environmental factors. Therefore, the durability of a piece of rock is an estimate of its ability to withstand environmental factors and affect its weathering, which depends on the mineralogical composition of the rock, the size and shape of its constituent minerals, the strength of the rock, the presence duration of the rock against the environmental factors and the chemical composition of environmental solutions (Fereidooni et al. 2015, 2016a, 2016b; Fereidooni 2016). Also, this feature represents the ability of different rocks to be decayed by chemical decomposition and their breaking and fragmentation physically under the influence of factors such as exfoliating, hydration, dissolution, oxidation and abrasion (Gupta and Ahmed 2007), which is insensitively affected by the mineralogical composition and the physical and mechanical properties of the rocks (Fereidooni and Khajevand 2018a). Until now, many researchers have focused on determining the durability and deterioration of rocks and the factors affecting them and have investigated this issue with various approaches and laboratory test methods. In the following, the most important laboratory test methods and indicators used to determine the durability and deterioration of rocks will be explained.

5.2 Slake-durability test

The slake-durability test has been widely used to evaluate the physical changes and slaking behavior of rocks as a result of the wetting and drying process. This test was firstly proposed by Franklin and Chandra (1972), and then its performing method was proposed by the International Society of Rock Mechanics (ISRM 1981) and the American Society for Testing and Materials (ASTM 1990), and so far, it has been successfully used on various rock types. However, there are some limitations and weaknesses related to this test method. To perform the test in

DOI: 10.1201/9781003672920-5

accordance with the above-mentioned standards, first, ten pieces of 40–60 g of each rock sample were prepared in which they did not have sharp corners. Then, these specimens are rotated in the test device, which contains a water container and a rotating mesh cylinder, for ten minutes at a speed of 20 rounds per minute (Figure 5.1). The diameter of the mesh cylinder holes is 2 mm. As a result of proximity to the solution inside the container, the rock pieces are destroyed due to contact with the cylinder wall and each other and the sediment particles due to their destruction deposited at the bottom of the container. After the test, the specimens are dried in the oven for 24 hours at a temperature of 105°C, and their weight loss is evaluated. Based on the standard, this test is performed in two cycles, and the remaining weight of each specimen after each cycle of the test to its initial weight is considered as the slake-durability index (Id). The temperature of the test solution should be 20°C, because the temperature and chemical composition of the solution used in the test affect the durability of the specimen. Based on the rock classifications proposed by Gamble (1971) and Franklin and Chandra (1972), different rocks are classified according to the slake-durability index, as shown in Tables 5.1 and 5.2. The value of the slake-durability index can be calculated from the following equation:

$$\mathrm{Id}_n = \frac{W_{dn}}{W_{d0}} \times 100 \tag{5.1}$$

In this equation, W_{d0} is the initial dry weight of the specimen and W_{dn} is the dry weight of the specimen in the nth cycle of the test. Id_n is also the slake-durability index for the nth cycle of the test.

The slake-durability test was successfully performed on different rocks, especially soft types such as sulfate rocks including gypsum and anhydrite (Kayabali et al. 2006), limestone, travertine and marble (Fereidooni et al. 2018; Fereidooni 2020, Danaei and Fereidooni 2023; Singh et al. 2006), hornfels (Ghobadi and Fereidooni 2015), sandstone (Hashemi and Fereidooni 2018),

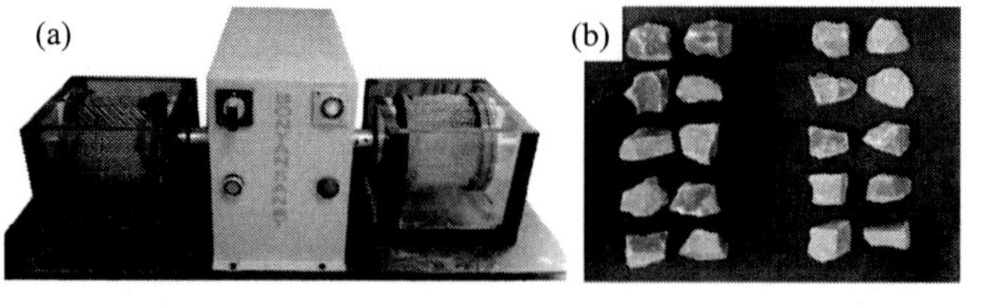

Figure 5.1 (a) The slake-durability test device and (b) the rock specimens before (left) and after (right) two cycles of test.

Table 5.1 Rock classification based on the slake-durability index (Gamble 1971)

Slake-durability index of the first cycle of the test (%)	*Rock durability description*
0–30	Very weak
30–60	Weak
60–85	Medium
85–95	Slightly durable
95–98	Durable
98–100	Very durable

Table 5.2 Rock classification based on the slake-durability index (Franklin and Chandra 1972)

Slake-durability index of the second cycle of the test (%)	*Rock durability description*
0–25	Very weak
25–50	Weak
50–75	Medium
75–90	Slightly durable
90–95	Durable
95–100	Very durable

tuff (Hashemi and Fereidooni 2023) and laterite (Sunil et al. 2006), and the relationship between the slake-durability index and other rock engineering properties has been investigated. Sharma and Singh (2008) investigated relationships between the slake-durability index and uniaxial compressive strength, Schmidt hardness and P-wave velocity of different rocks, including sandstone, basalt, mica schist, coal and shale. Yagiz (2011), in a complete study, detailed experimental relationships between slake-durability index and uniaxial compressive strength, Schmidt rebound hardness, wave velocity, modulus of elasticity, effective porosity, water absorption and dry and saturation unit weights of seven types of carbonate rocks obtained from southwestern Turkey. Kahraman et al. (2016) estimated the uniaxial compressive strength with high accuracy through the slake-durability index of pyroclastic rocks. Some researchers (e.g. Fereidooni 2016; Fereidooni and Khajevand 2018b) have also tried to establish relationships between the slake-durability index with other physical and mechanical properties of rocks using regression analysis, artificial neural network and fuzzy logic methods and found that these parameters are completely related to each other.

Most previous research studies investigating the slake-durability of rocks have been based on performing two cycles of the slake-durability test. But Ulusay et al. (1995) used the results of the slake-durability test based on five cycles to evaluate the slake-durability of marl columns of a coal mine. They concluded that by increasing the number of test cycles, a significant decrease in the slake-durability index of the rocks is observed. Also, by examining the granularity of the particles passing through the mesh cylinder of the test device, they found that with the increase in the size of the specimens, the percentage of the particles increased after performing five cycles of the test. Bell et al. (1997) studied the tailings of several coal mines in the United Kingdom and stated that a three-cycle test is the best method to evaluate the durability of sandstone and mudstone. Gokceoglu et al. (2000) assessed the factors affecting the slake-durability index of clay-bearing weak rocks, together with the effect of wetting and drying cycles on them, with attention to the presence of expanding clay minerals. They reported that the best relationship between the slake-durability index and the uniaxial compressive strength of the rocks is after the fourth cycle. According to Fereidooni (2016), any increase in the number of test cycles reduces the decreasing rate of the slake-durability index. This means that this reduction of the index is more evident in the initial cycles of the test.

Other researchers indicated that the durability of different rocks is highly dependent on their petrographic characteristics. In this regard, Tugrul and Zarif (1999) stated that the mineralogical composition of sandstones has the greatest control over the rock's durability. In other words, the texture and fabric characteristics of the sandstones are more important effective parameters on their durability than the mineralogy of the grains. Dhakal et al. (2002) observed that mineralogy has a significant relationship with the slake-durability index and weathering degree of rocks. Gupta and Ahmed (2007) stated that fine-grained limestones are more susceptible to decomposition than coarse-grained ones and other carbonate rocks, especially in acidic environments. It has also been stated that the mineralogical composition, physical properties and rock texture (degree of locking of crystals, shape and size of crystals, surface roughness, crystal surface area, crystal perimeter and effective porosity) are mainly related to the slake-durability index of rocks and affect their mechanical properties (Yilmaz and Karacan 2005; Kayabali et al. 2006; Ghobadi and Fereidooni 2015).

Applying wetting and drying cycles during the slake-durability test indicates that this phenomenon is a very effective factor in the process of rock slaking. According to Khajevand and Fereidooni (2022), if the pH of the fluid solution of the slake-durability test decreases, the slake-durability index will decrease, and this issue is more evident in the initial cycles of the test. Singh et al. (2006) investigated the effect of acidic water on the slake-durability index of shales in laboratory conditions using artificial neural networks and fuzzy logic methods to predict the slake-durability index of the rocks.

In the study of Danaei (2021), eight rock samples of common building stones of Iran were selected. To determine the slake-durability index of the rocks, specimens with the shapes of lump, rounded lump, cube, sphere, square prism, square-base plate and rectangular cube were tested in the laboratory. The main purpose of this research was to investigate the effect of specimen shape on the slake-durability index of the rocks after ten cycles of the slake-durability test and also to investigate the grain size distribution of the sediment particles obtained from the test. After performing ten cycles of the tests on the rock specimens, it was found that the spherical specimens have the highest durability and the lump specimens have the lowest value of the slake-durability index due to sharp angles and corners. Therefore, whatever the specimens have regular dimensions and no sharp corners, they are more durable. In this research, after the slake-durability test, the obtained sediment particles from the test were granulated using the sieving technique, and sediment granulation parameters, including D_{10}, D_{30} and D_{60}, were evaluated. Then, the uniformity and curvature coefficients were calculated for the sediment particles obtained from the specimens. According to the unified classification system, most of the sediment particles obtained from the slake-durability test were placed in the ML class. The granulation parameters of these soils were different and completely affected by the specimen shapes of the slake-durability test.

Based on previous research studies on the determination of rock durability using the slake-durability test, it is evident that numerous factors have been identified as influential. These factors can be broadly classified into three groups, including (i) rock-related factors, (ii) test fluid-related factors or environmental factors and (iii) laboratory factors. Key factors among these include the porosity and permeability of the rock, the nature (temperature, pH and composition) of the test fluid, the rock stability against swelling and decomposition, the size and weight of specimens, the characteristics of the testing equipment, the roundness and shape of the specimens, the number of wetting and drying cycles and finally the competence of the experimenter.

5.3 Freezing and thawing test

The stones used in the construction of historical monuments and building stones used in cold regions are yearly exposed to freezing and thawing natural cycles. In tropical regions, in the seasons of the year when the temperature drops below zero in the middle of the night, there is also a possibility of such a phenomenon. The process of repeated freezing and thawing causes a decrease in strength and the gradual decomposition and destruction of rocks and changes their physical and mechanical properties. The amount of decomposition and destruction caused by the process of freezing and thawing in rocks is very important in road construction projects, railways, pipelines and construction materials. Also, repeated freezing and thawing are among the most important processes of mechanical weathering. The process of destruction of rocks due to freezing and thawing is such that during the freezing time, the water in the small voids of rock (with less than 10 μm in dimension) increases in volume by about 9%, which leads to the occurrence of tensile stress, and then micro-fractures in the rock (Gokce et al. 2016). Also, there is a possibility of old micro-fractures developing following the increase in the tensile stress due to the freezing and thawing process in the rock (Chen et al. 2004). During the time of ice thawing, water flows through the destroyed micro-fractures and increases the amount of destruction. Also, the freezing of water in the empty spaces in the rock causes the formation of ice crystals, and this phenomenon causes internal stresses to be applied to the constituent particles of the rock. The durability of rocks against successive freezing and thawing, as well as their durability against wetting and drying, depends on the degree of rock saturation, porosity, distribution and size of voids and also the range of environment temperature changes. Rocks with low permeability, in which water hardly flows, are more susceptible to destruction during freezing and thawing due to the existence of significant non-uniform stresses (Saad et al. 2010). Figure 5.2 shows a schematic view of water infiltration into the voids and cavities of rock, which destroys the rock by freezing and increasing its volume.

The freezing and thawing test focuses on simulating the stresses created in rocks during the formation of ice crystals in its voids and assessing the durability of the rock and is generally controlled by changes in temperatures below and above 0°C. This test is performed in accordance with the ASTM-D5312 (2013) standard. To perform the test, rock specimens with different shapes such as lumpy, cubic, cylindrical- and disc-shaped cores with a diameter of NX are saturated in water. After this step, the rock specimens are placed in a freezer at −18°C for 15 hours so that the water that has penetrated into their voids freezes and increases in volume. Then the specimens are immersed in a water tank with a temperature of 32°C for 9 hours so that the frozen water in the rock voids thaws again. Electric heaters are used to keep the water temperature constant. This test is performed according to the standard in five cycles, although some researchers have also performed this test in a higher number of cycles. After the completion of the test cycles, the changes in the physical and mechanical properties and weight loss of the specimens are checked by calculating indices in the form of the percentage drop of these parameters to their initial values for the rock specimens. After performing each cycle of the freezing and thawing test, the amount of weight loss, the reduction of weight-volume indices and the loss of mechanical properties are calculated as a percentage of the initial values of these parameters for the rock specimens. Figure 5.3 illustrates the performing process of the freezing and thawing test and its repetition cycles.

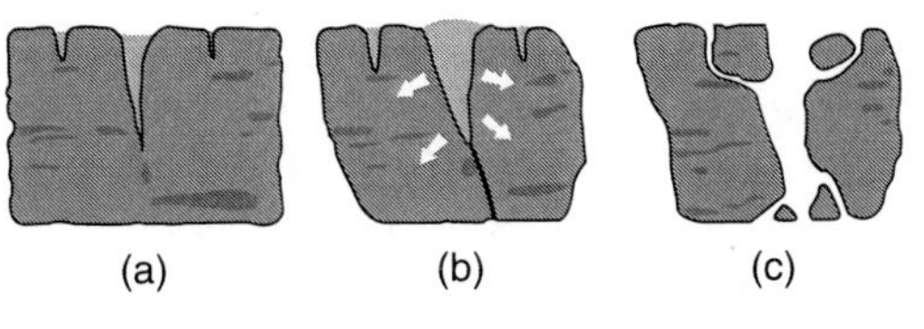

Figure 5.2 Mechanism of freezing, volume increasing and destruction of rock: (a) filling of rock fractures and voids with water, (b) freezing the water and providing compressive force to the inner surface of the fractures and (c) destruction of rock structure due to the opening and expansion of the fractures.

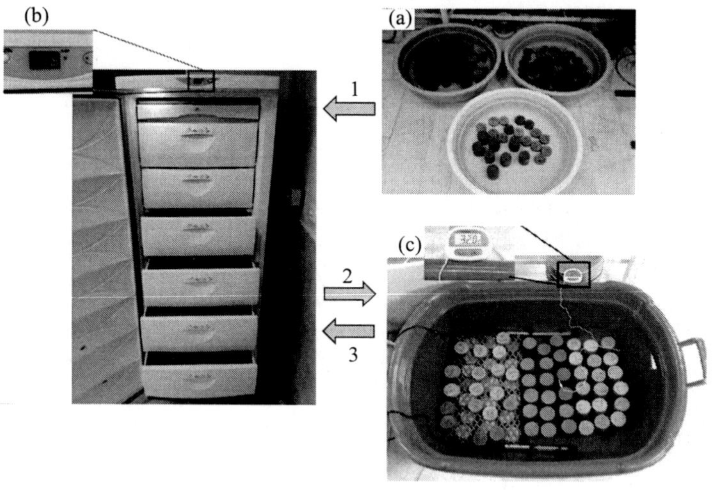

Figure 5.3 Freezing and thawing test process and repetition of its cycles (adapted from Hashemi 2018): (a) saturating specimens by immersing them in water, (b) placing the specimens in the freezer to cause freezing and (c) immersing the specimens in water to cause thawing.

5.4 Thermal shock test

Seasonal or daily temperature changes cause the thermal expansion and contraction of rocks, so that due to thermal fluctuations, new cracks are created or their existing cracks develop. In general, the coefficient of volume expansion for rocks varies between 15×10^{-6} and 33×10^{-6} (Steiger et al. 2011). It is obvious that with the increase of this coefficient, the amount of expansion and contraction of rocks increases, and the damage caused to it due to the temperature changes increases. Also, the rocks composed of several minerals show a great variation in the expansion rate due to the difference in the thermal expansion coefficients of their minerals. So, if the temperature fluctuation and the repetition of these conditions in the rocks reach enough, destruction will occur in the rocks. The degradation of rocks due to temperature change is significant when water is present in the environment. The igneous rocks, especially coarse crystalline types that are composed of several different minerals, are more affected by temperature changes than other rocks. Thermal shock often occurs in areas with hot and desert climates that have hot long days and cold nights (Yavuz 2006). Therefore, with the increase in temperature, physical and mechanical changes occur in the rock, and after the rock cools down, most of these changes are irreversible. The difference in the thermal expansion properties of the minerals composing the rock can cause great stresses within the rock.

The thermal shock test simulates the environmental temperature changes in the laboratory. In this test, the rock specimen is exposed to temperature fluctuations. The thermal shock test is performed in accordance with BS EN 14066 (2013). In this test, specimens with a regular geometric shape such as cylindrical and disc-shaped rock cores with a diameter of NX are suitable. The heating stage in this test consists of placing the specimens in an oven at a temperature of 105°C for 18 hours, and the cooling stage (shocking) consists of submerging the specimens for 6 hours in a water tank with a temperature of 20°C. Therefore, each cycle of this test takes 24 hours. After performing the test cycles, the physical and mechanical properties and weight loss percentage of the specimens are calculated and expressed as a percentage of the initial

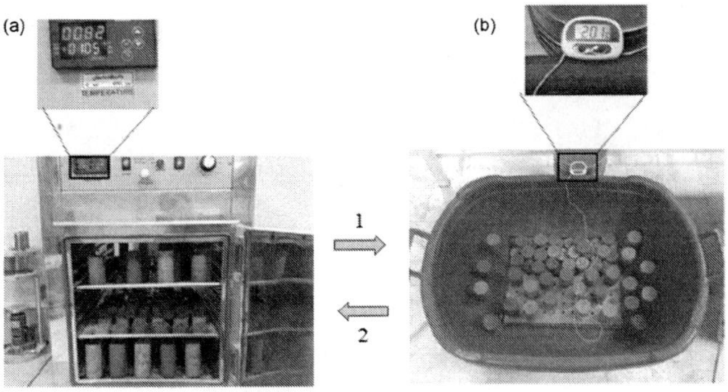

Figure 5.4 Thermal shock test process and repetition of heating and cooling cycles (adapted from Hashemi 2018): (a) placing specimens in the oven at a temperature of 105°C and (b) immersing the specimens in water with a temperature of 20°C.

values of these parameters. Figure 5.4 illustrates the performance performing process of the thermal shock test and the repetition of its heating and cooling cycles.

5.5 Salt crystallization test

Weathering caused by salt crystallization is one of the strongest types of weathering in rocks, which changes the engineering properties of the rock and reduces its durability. Studying the mechanism of salt crystallization and understanding its destructive effects on rocks is very important to find an approach to prevent or limit this phenomenon in the future. Common salts in rocks and building stones generally originate from atmospheric precipitation, underground water and street and road waters. In this process, the solutions containing salt dissolve in contact with the rock materials and penetrate into the rock voids and cavities, and the crystallization of dissolved salts in them expands during the evaporation phase and causes destruction of the rock from the inside. The crystallization pressure in large cavities is usually low, while in small fully saturated voids, high crystallization pressures are expected. In this regard, when the pressure of salt crystallization exceeds the tensile strength of the rock, it will be damaged.

Among the types of salts available in nature, we can mention sulfate salts. The sulfates are a combination of sulfur and oxygen and are found as a part of natural minerals in the composition of some soils and rocks. The sulfate salts are sodium and magnesium sulfates. The occurrence of sodium and magnesium sulfate salts is completely regular in nature, and their distribution has been reported all over the world (Ghoudie and Cooke 1984). Sodium sulfate is found in nature in anhydrous form (thenardite) and aqueous form (mirabilite). When thenardite transforms into mirabilite, an increase in volume occurs, which leads to salt pressure. The sodium sulfate solution has more destructive effects than other salts.

BS EN 12370 (1999) standard is used for performing the salt crystallization test. For this purpose, firstly, rock specimens with a regular geometric shape such as cylindrical and disk-shaped specimens with a diameter of NX are provided and they are immersed in a salt solution such as sodium sulfate of 14% by weight (14 g of sodium sulfate in 86 g of water) at a temperature of 20°C for 2 hours. Then, after removing and drying the surface of the specimens, they are placed in an oven at a temperature of 105°C for 15 hours. In the last stage, the specimens are cooled for 7 hours at the laboratory temperature and the next test cycle begins again. Therefore, it takes 24 hours to complete each cycle of the test. After performing the test cycles, the physical,

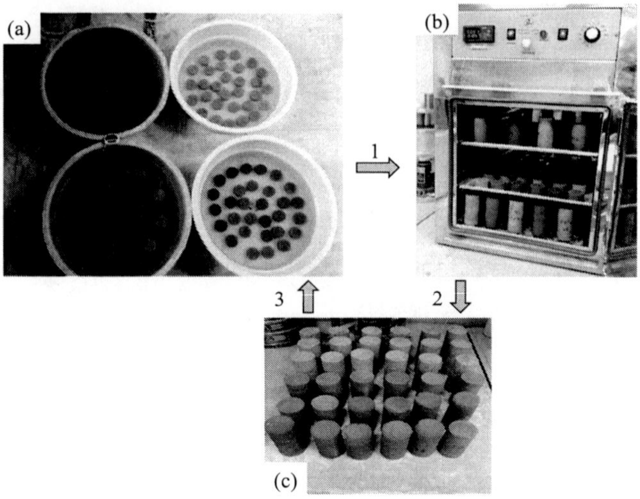

Figure 5.5 Salt crystallization test process (adapted from Hashemi 2018): (a) immersing and saturating specimens in salt water solution, (b) placing the specimens in an oven to create salt crystallization and (c) cooling the specimens and re-immersing them in the saltwater solution.

mechanical and weight loss characteristics of the specimens are evaluated and expressed as a percentage of the initial values of these parameters. Figure 5.5 illustrates the performing process of the salt crystallization test.

5.6 Accelerated weathering test caused by acid rain

The development of cities, their industrialization and the use of fossil fuels are associated with the creation of acid rain with a pH of less than 5, which always affects building stones. Water is one of the main factors of weathering. Due to the increase of dissolution ability of minerals and rocks with increasing the acidity degree of acidic solutions, acid rains cause more weathering in the rocks than normal rains (Fereidooni and Ghasemi 2025). The destruction of historical monuments and stone facades of buildings in industrial areas confirms this issue. Due to the existing limitations in the standard slake-durability test method, the existence of distinctions compared to the natural weathering conditions of rocks under the influence of acid rains, and the creation of multi-purpose conditions for the simultaneous examination of different parameters, Fereidooni and Khajevand (2019) in the Damghan University built an accelerated weathering test device with the ability to produce natural and acid rains. The explanation of how the device works and its results as a new method to determine the durability and deterioration of rocks are assessed in this part.

5.6.1 Amis of accelerated weathering test

The purpose of performing the accelerated weathering test is to create natural weathering conditions, to investigate the effect of natural and acidic rainfalls only during wetting and drying cycles on durability, weight loss rate, changes in the physical characteristics (including specific gravity, dry and saturated unit weights, porosity and water absorption) and changes in the index and mechanical characteristics of rocks (such as ultrasonic wave velocity, uniaxial compressive strength, Brazilian tensile strength and block punch strength). The accelerated weathering tester generally simulates a cycle of normal and acidic precipitation on rock specimens. In this method, by placing the rock specimens inside the encasement of the device and causing rainfall, simultaneously and during each cycle, the change in the above-mentioned parameters under acidic and normal precipitations can be evaluated and compared.

5.6.2 Description of accelerated weathering test device

The accelerated weathering test device was built after determining the goals and perspective of the studies. Figure 5.6 shows a general image of the accelerated weathering test device. The main encasement of the device, which is the location of rock specimens for testing, has dimensions of $90 \times 50 \times 90$ cm (length, width, height) and a net with 2-inch grids is placed on its bottom. It is an aquarium acid tank with dimensions of $40 \times 40 \times 70$ cm, which is made of transparent glass with a thickness of 6 mm and has a capacity of 76 L of solution. It is placed at the bottom of the main encasement. The acidic solution is poured on the specimens after being sucked by the electro pump and leaving the precipitation outlets, and then it returns to the solution reservoir from the discharge point, and this process is cyclically repeated. The used electro pump is a 220V electric dynamo device with a power of 0.5 horsepower and 0.37 KW, which is equipped with a pump plated with a galvanized copper layer and is resistant to acids. The main skeleton of the device is made of steel profile, in which, to prevent rust and decay, first the pores in the joints are filled with Mel powder, and then its surface is covered with a layer of stainless acrylic coating and finally with oil resin. To insulate the main encasement and prevent water leakage, 8 mm thick transparent polycarbonate plexiglass and colorless silicone glue have been used. The pipes and fittings used to prevent corrosion by the acid solution used are five-layer UPVC type without glue, which are highly resistant to all forms of corrosion due to the non-metallic nature of these materials. Control valves are lever-operated and made of brass, and they are sealed with polyethylene O-rings at the tongue. The power transmission cables have double-layer copper conductors with a non-flammable coating made of polyvinyl chloride with a size of 2.5 and are covered with aluminum waterproof insulation in the internal parts of the device.

To control as much as possible the amount of water and acidic solutions used in the test, control valves are installed on each inlet and outlet. To facilitate the production process of carbonic

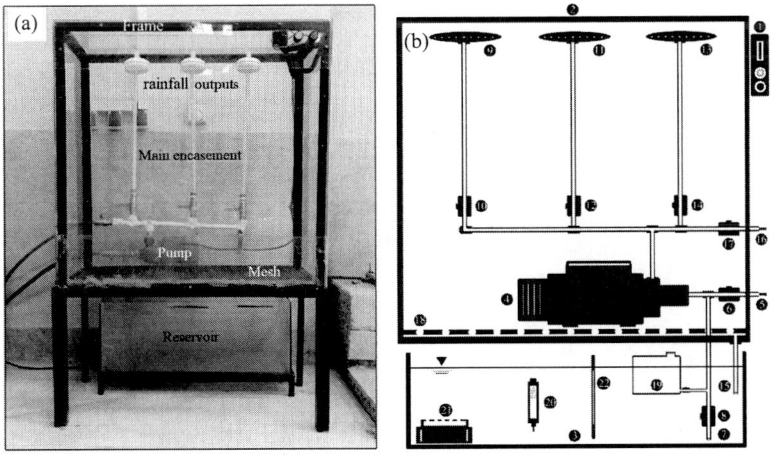

Figure 5.6 (a) Image and (b) technical map of accelerated weathering test device (Fereidooni and Khajevand 2019).

acid, if needed, a place is provided to install the CO_2 capsule. When the device starts raining, it starts and continues by entering the aeration tank in the circuit as a driving factor for the circulation of the solution.

5.6.3 *Accelerated weathering test performing method*

After specimen preparation, a set of specimens, which include five cylindrical, disk, or tablet-shaped specimens, are placed on the chassis of the specimen location, as shown in Figure 5.7. Then, the specimens are subjected to rain with natural water during a cycle for 10 minutes and with a flow rate of 30 L/min. at a temperature of 20°C. This cycle can be performed under similar conditions for acidic solutions such as sulfuric or nitric acids with different pHs. After performing each cycle of the test, the specimens are placed in the oven at a temperature of 105°C for 24 hours. The ratio of the dry weight of each specimen at the end of each cycle to its initial dry weight before the start of the test is considered the accelerated weathering index for that specimen. In the following relation, the general formula for calculating the accelerated weathering index is presented:

$$AWI_n = \frac{W_{dn}}{W_{d0}} \times 100 \tag{5.2}$$

In this equation, W_{d0} is the initial dry weight of the specimen and W_{dn} is the dry weight of the specimen in the nth test cycle. AWI_n is also the accelerated weathering index for the nth test cycle. This index is calculated as a percentage for all test cycles. This test can be performed in a

Figure 5.7 Accelerated weathering test process performed on cylindrical, disc and thin disc-shaped specimens under natural water and acidic solution rainfalls.

large number of cycles. But it is better to do it at least in four cycles. The results of performing this test in two cycles are comparable to the results of the standard slake-durability test. After performing the accelerated weathering test cycles, other rock engineering characteristics such as physical and mechanical properties can be calculated and expressed as a percentage of the initial values of these parameters.

References

ASTM, 1990. *Standard Test Method for Slake Durability of Shales and Similar Weak Rocks. ASTM-D-4644. Annual Book of ASTM Standards*. ASTM International, West Conshohocken, PA, Vol. 4.08, pp. 863–865.

ASTM, 2013. *Standard Test Method for Evaluation of Durability of Rock for Erosion Control under Freezing and Thawing Conditions. D5312, ASTM standards*. ASTM International, West Conshohocken, PA.

Bell, F.G., Entwisie, D., Culshaw, M., 1997. A geotechnical survey of some British coal measures mudstones, with particular emphasis on durability. *Engineering Geology*, 46(2), 115–129.

BS EN 12370, 1999. *Natural Stone Test Methods—Determination of Resistance to Salt Crystallization*. British Standards Institute, London.

BS EN 14066, 2013. *Natural Stone Test Methods, Determination of Resistance to Ageing by Thermal Shock*. British Standards Institute, London.

Chen, T.C., Yeung, M.R., Mori, N., 2004. Effect of water saturation on deterioration of welded tuff due to freeze-thaw action. *Cold Regions Science and Technology*, 38, 127–136.

Danaei, S., 2021. The effect of specimen's shape on the slake-durability index of rocks. Master's Thesis in Geology (Engineering Geology), Damghan University, Damghan, Iran, 126 pages.

Danaei, S., Fereidooni, D., 2023. On the importance of specimen's geometric shape effects on the slake-durability index of limestones and grain size distribution of the sediment particles obtained during the test. *Construction and Building Materials*, 394, 132205.

Dhakal, G., Yoneda, T., Kato, M., Kaneko, K., 2002. Slake durability and mineralogical properties of some pyroclastic and sedimentary rocks. *Engineering Geology*, 65(1), 31–45.

Fereidooni, D., 2016. Determination of the geotechnical characteristics of hornfelsic rocks with a particular emphasis on the correlation between physical and mechanical properties. *Journal of Rock Mechanics and Geotechnical Engineering*, 49, 2595–2608.

Fereidooni, D., 2020. Reply to "Comments on: Correlations between slake-durability index and engineering properties of some travertine samples under wetting–drying cycles" by Jamshidi and Sarikhani. *Geotechnical and Geological Engineering*, 38, 1013–1015. https://doi.org/10.1007/s10706-019-01008-7.

Fereidooni, D., Ghasemi, F., 2025. Impact of simulated rainfall on engineering properties of carbonate building stones. *Geotechnical and Geological Engineering*, 43, 52.

Fereidooni, D., Hashemi, M., Sabouri, V., 2018. Assessing the correlation between slake-durability index and Los Angeles abbreviation for limestone aggregates. *International Conference on Civil Engineering, Architecture and Urban Management*. University of Tehran, Tehran, Iran.

Fereidooni, D., Khajevand, R., 2018a. Determining the geotechnical characteristics of some sedimentary rocks from Iran with an emphasis on the correlations between physical, index, and mechanical properties. *Geotechnical Testing Journal*, 41(3), 555–573.

Fereidooni, D., Khajevand, R., 2018b. Correlations between slake-durability index and engineering properties of some travertine samples under wetting–drying cycles. *Geotechnical and Geological Engineering*, 36, 1071–1089.

Fereidooni, D., Khajevand, R., 2019. Utilization of the accelerated weathering test method for evaluating the durability of sedimentary rocks. *Bulletin of Engineering Geology and the Environment*, 78, 2697–2716.

Fereidooni, D., Khajevand, R., Fadakar, H., 2016a. Assessment of durability and influence of acidic rainfalls on weathering and degradation of Karaj Formation tuffs in Damghan area. *Proceedings of the 34th National and the 2nd International Geosciences Congress*. March 3–5, Tehran, Iran.

Fereidooni, D., Khajevand, R., Fathi, P., 2015. Assessment of degradation and weathering of limestone samples under acidic rainfalls. *Proceedings of the Second National Conference on Soil Mechanics and Foundation Engineering*, Qom University of Technology, October 22 and 23, Qom, Iran.

Fereidooni, D., Tabatabai Jafari, S.A., Ghasemi, A., 2016b. Assessment of acidic rainfall influences on degradation and weathering of sandstones from Shemshak Formation in northwest of Damghan, Iran (in Persian). *Proceedings of the 34th National and the 2nd International Geosciences Congress*, March 3–5, Tehran, Iran.

Franklin, J.A., Chandra, R., 1972. The slake-durability test. *International Journal of Rock Mechanics and Mining Science*, 9, 325–341.

Gamble, J.C., 1971. Durability-plasticity classification of shales and other argillaceous rock. University of Illinois, USA, Ph.D. Thesis, 159 pages. https://hdl.handle.net/2142/64715.

Ghobadi, M.H., Fereidooni, D., 2015. Effect of mineralogy on durability and strength of hornfelsic rocks under acidic rainfall in urban areas. *Journal of Engineering Geology*, 9(2), 2765–2788.

Ghoudie, A.S., Cooke, R.U., 1984. Salt efflorescences and Saline lakes; a distributional analysis. *Journal of Geoforum*, 15(4), 563–582.

Gokce, M.V., Ince, I., Fener, M., Taskiran, T., Kayabali, K., 2016. The effects of freeze–thaw (F–T) cycles on the Gödene travertine used in historical structures in Konya (Turkey). *Journal of Cold Regions Science and Technology*, 127, 65–75.

Gokceoglu, C., Ulusay, R., Sonmez, H., 2000. Factors affecting the durability of selected weak and clay-bearing rocks from Turkey, with particular emphasize on the influence of then number of drying and wetting cycle. *Engineering Geology*, 57, 215–237.

Gupta, V., Ahmed, I., 2007. Effect of pH of water and mineralogical properties on the slake durability (degradability) of different rock From the Lesser Himalaya India. *Engineering Geology*, 95, 87–79.

Hashemi, M., 2018. Assessing the engineering properties of the rocks employed in historical stone monuments of southern Zanjan province with an emphasis on durability. Master's Thesis in Geology (Engineering Geology), Damghan University, Damghan, Iran, 149 pages.

Hashemi, M., Fereidooni, D., 2018. Slake-durability assessment of rocks from the historical stone monuments of southern Zanjan province. *5th National conference on Applied Research in Civil Engineering, Architecture and Urban Management*, January 13–14, K.N. Toosi University of Technology, Tehran, Iran.

Hashemi, M., Fereidooni, D., 2023. Investigating the effect of weathering on the durability and deterioration of historical stone monuments in the south of Zanjan province. *Journal of Engineering Geology*, 17(2), 187–203.

ISRM, 1981. Rock characterization testing and monitoring. In: Brown E.T. (ed) Pergamon Press, Oxford.

Kahraman, S., Fener, M., Gunaydin, O., 2016. Estimating the uniaxial compressive strength of pyroclastic rocks from the slake durability index. *Bulletin of Engineering Geology and the Environment*, 76, 1107–1115.

Kayabali, K., Beyaz, T., Kolay, E., 2006. The effect of the pH of the testing liquid on the slake durability of gypsum. *Bulletin Engineering Geology and the Environment*, 65, 65–71.

Khajevand, R., Fereidooni, D., 2022. The effects of water acidity and engineering properties on rock durability. *Earth Sciences Research Journal*, 26(1), 67–79.

Saad, A., Guedon, S., Martineau, F., 2010. Microstructural weathering of sedimentary rocks by freeze–thaw cycles: Experimental study of state and transfer parameters. *Competes Rendus Geosci,* 342:197–203.

Sharma, P., Singh, T., 2008. A correlation between P-wave velocity, impact strength index, slake durability index and uniaxial compressive strength. *Bulletin of Engineering Geology and the Environment*, 67, 17–22.

Singh, T., Sharma, P., Khandelwal, M., 2006. Effect of pH on the physico-mechanical properties of marble. *Bulletin of Engineering Geology and the Environment*, 66(1), 81–87.

Steiger, M., Charola, AE., Sterflinger, K., 2011. Weathering and Deterioration. In: Siegesmund, S., Snethlage, R., (eds), *Stone in Architecture: Properties, Durability*, (4th edn, pp. 227–316). Springer, Berlin.

Sunil, B.M., Nayak, S., Shrihari, S., 2006. Effect of pH on the geotechnical properties of laterite. *Engineering Geology*, 85, 197–203.

Tugrul, A., Zarif, I., 1999. Correlation of mineralogical and textural characteristics with engineering properties of selected granitic rocks from Turkey. *Engineering Geology*, 51, 303–317.

Ulusay, R., Arikan, F., Yoleri, M.F., Caglan, D., 1995. Engineering geological characterization of coal mine waste material and an evaluation in the context of back-analysis of spoil pile instabilities in a strip mine SW Turkey. *Engineering Geology*, 40, 77–101.

Yagiz, S., 2011. Correlation between slake durability and rock properties for some carbonate rocks. *Bulletin of Engineering Geology and the Environment*, 70(3), 377–383.

Yavuz, A.B., 2006. Deterioration of the volcanic kerb and pavement stones in a humid environment in the city centre of Izmir, Turkey. *Environmental Geology*, 51(2), 211–227.

Yilmaz, I., Karacan, E., 2005. Slaking durability and its effect on the doline formation in the gypsum. *Environmental Geology*, 47, 1010–1016.

Chapter 6

Hardness and abrasiveness

6.1 Introduction

Rocks have been used by humans as one of the most important building materials everywhere in the world throughout history. One of the most important features of rocks as building materials is their surface hardness and abrability, especially when they are used as floor stones and pavements. The surface hardness of substances is determined by various tests in engineering and material sciences. To measure the rock/stone surface hardness in the laboratory, we are always faced with two common words of *abrasiveness* and *abrasivity*. In terms of nomenclature, they are very similar and often confused. Even within the academic community, researchers often use the names interchangeably, which needs to be corrected. It should be emphasized here that abrasiveness is determined when we are interested in the abrasive wear of natural stone and abrasivity when we are interested in the wear of the tool with which we cut the stone (Mucha 2023).

The main purpose of different methods for determining the rock surface hardness is to estimate its strength, durability, cutting and drilling ability and wear resistance to determine its quality and suitability in civil constructions and the ability of underground excavations. The concept of surface hardness is the stability of a material against crushing, scratching, penetration of a harder material or the amount of impact an object has when it hits the rock surface. In rocks, this characteristic depends on rock type, constituent minerals, the binding strength of minerals as well as the degree of porosity and moisture of the rock (Diamantis and Fereidooni 2023). Different hardness tests were initially designed and used to determine the surface hardness of metals that have a smooth and homogeneous surface. Unlike metals, which have a smooth and homogeneous surface, rocks have an uneven and heterogeneous surface. Therefore, for rocks that have heterogeneous surface and coarse structures, better results will be obtained by using a larger indenter or rebounding object and applying more force.

The simplest types of hardness testing methods include crushing the rock with a hammer or scratching its surface with a rasp. In these methods, the shards and scratches created will quantitatively and qualitatively indicate the relative hardness of the rock surface. The amount of crushing or abrasion of the rock surface under the specified conditions is called the hardness of crushing and abrasiveness or resistance to crushing and abrasiveness, respectively. If a certain substance is dropped from a certain height on the rock surface (which has elastic properties) and the resistance of that substance is measured, this test is called the rebound hardness test. But the most common hardness test is the penetration of a penetrating or indenting object into the substance subjected to the test and recording the force required for this action or measuring the amount of indentation against a certain force. This type of test is called indentation hardness test. In this case, the numerical value of hardness can be determined in two ways, including (i)

DOI: 10.1201/9781003672920-6

determining the depth of the penetrating object and (ii) determining the size of the penetrating object's impression surface. Determining the surface hardness of rocks is one of the simple, fast and cheap methods that can be used as an index test such as point load, block and cylinder punch, wave velocity and slake-durability tests to estimate the physical and mechanical properties of the rocks (Fereidooni 2016; Fereidooni et al. 2016; Khanlari et al. 2014a, 2014b; Fereidooni and Khajevand 2018; Khajevand and Fereidooni 2018; Fereidooni et al. 2024).

6.2 Rock hardness determination methods

Hardness determination methods are different based on basics, logic, mechanism, scale and time. Depending on how the hardness test is performed, the existing tests can be divided into two general categories, including static tests and dynamic tests (Fereidooni 2023).

6.2.1 *Static hardness tests*

In static hardness determination methods, the effect of an object on a rock surface is checked by applying a static force. This method includes crushing hardness, scratch hardness and indentation hardness. The crushing hardness is measured using the Rosiwal hardness test, the scratch hardness is measured using the Mohs hardness testing method, and the indentation hardness includes Vickers, Knoop, Rockwell hardness, cone indentation, nanoindentation tests and Cerchar abrasivity index. Among the above three groups of methods, the indentation tests are considered the most important hardness determination methods in engineering sciences. In these tests, which are also known as penetration hardness tests, the hardness is measured by the surface resistance of the substance against the indentation. In these methods, by penetrating an indenter (ball, pyramid or hard steel cones) and impressing the substance, the hardness of its surface is measured based on the depth and dimensions of the resulting impression (Ghasemi and Fereidooni 2019).

6.2.1.1 *Mohs hardness*

The Mohs hardness test is one of the most important, accurate and fundamental methods for measuring the hardness of minerals, which was introduced by Friedrich Mohs. This test compares the resistance of a mineral to being scratched by ten reference minerals known as the Mohs scale of hardness. Reference hardness is assigned to talc, gypsum, calcite, fluorite, apatite, orthoclase, quartz, topaz, corundum and diamond, in order from one to ten. This scale is also the most popular and practical method for evaluating and classifying rock hardness, because it is directly based on mineralogical studies and has a very good ability to analyze rock hardness (Fereidooni 2022a). Minerals of the Mohs hardness scale and their characteristics are listed in Table 6.1. It should be noted that sometimes other substances such as a human nail (hardness of 2.5), knife blade (hardness of 4.5) and glass (hardness of 5.5) are used to determine the hardness with the mouse method.

It is important to know that the Mohs scale of hardness considers the different orientations of crystals in minerals. This difference is due to the atomic structures and different bonds in minerals. Therefore, to accurately determine the hardness of a rock composed of several minerals, the percentage of minerals in each part of the rock is determined. Then, based on the hardness of each mineral, the whole rock hardness is determined by using the following equation (Fereidooni 2022b):

$$\mathrm{MH} = \sum_{i=1}^{n} A_i \times (\mathrm{MH})_i \tag{6.1}$$

Table 6.1 Mohs hardness scale and its reference minerals

Hardness	*Mineral*	*Chemical formula*	*Crystal sample*
1	Talc	$Mg_3(Si_4O_{10})(OH)_2$	
2	Gypsum	$CaSO_4.2H_2O$	
3	Calcite	$CaCO_3$	
4	Fluorite	CaF_2	
5	Apatite	$Ca_5(PO_4)_3(F,OH,Cl)$	
6	Orthoclase	$KAlSi_3O_8$	
7	Quartz	SiO_2	
8	Topaz	$Al_2(SiO_4)(F,OH)_2$	
9	Corundum	Al_2O_3	
10	Diamond	C	

In this equation, MH is the rock hardness, A_i is the percentage of each mineral in the rock and $(\text{MH})_i$ is the hardness of the mineral.

6.2.1.2 *Rosiwal hardness*

The Rosiwal hardness is considered a measure of the resistance of a rock or mineral under the abrasivity of an abrasive object. This scale is based on the reaction of minerals against a standard abrasive powder that is used in small amounts in preliminary tests. The used powder is corundum powder or, in some cases, dolomite or quartz powder, whose efficiency is calibrated against corundum powder. In this test, a specimen with an area of 400 mm^2 is pressed by hand on a rotating metal or glass disk until the powder loses its properties. Then, the milling time of 8 minutes was considered a standard and the amount of the used corundum powder was determined to be 100 mg (Fowell and Abu Bakar 2007). In other words, the specimen is abraded using a standard amount of abrasive powder until the powder wears out and reaches a constant weight. In this method, the weight loss of the specimen is a measure to determine its abrasiveness hardness. In the Rosiwal hardness scale, a hardness number of 1000 is assigned to corundum and the hardness of other minerals is measured accordingly. Similar to the methods of Mohs and Vickers, it is possible to obtain the total Rosiwal hardness by using a thin section of rock and the mineralogical description of its components according to the following equation (Yilmaz 2011):

$$\text{RoH} = \sum_{i=1}^{n} A_i \times (\text{RoH})_i \tag{6.2}$$

In this equation, RoH is the rock's Rosiwal hardness, A_i is the percentage of each mineral in the rock and $(\text{RoH})_i$ is the Rosiwal hardness of the mineral. According to the following relationship, the Rosiwal hardness of rocks is related to their Mohs hardness with good accuracy (R^2=0.95) (Thuro and Plinninger 2003):

$$\text{MH} = 1.05 \ln \text{RoH} + 2.12 \tag{6.3}$$

In this equation, RoH is the Rosiwal hardness of the rock and MH is its Mohs hardness. This relationship was obtained based on determining the hardness of 24 different minerals, except diamond.

6.2.1.3 *Brinell hardness*

One of the oldest methods of hardness measurement was introduced by Johan August Brinell, a Swedish engineer, in 1900 (Brinell 1900). In this method, a ball, which usually has a diameter of 10 mm, is used to apply force to the surface of the object. This force is applied for 10–30 S. The ball is made of steel or tungsten carbide. The applied force is different for various objects, for example, 3000 kg is applied to hard metals, 1500 kg to medium hard metals and 500 kg to soft materials. Brinell's hardness number is equal to the ratio of the force applied to the indentation effect area on the material surface. In this method (unlike the Rockwell method), the depth of the indentation effect is not considered and the diameter of the indentation effect is a measure of hardness. To get the exact diameter of the indentation, two intersecting perpendicular diameters are determined and averaged. The Brinell test can be performed by a universal testing machine by adding a suitable transducer to hold the ball. In the standard Brinell test, the diameter of the hole created by the ball is measured by a microscope, micrometer or microscope with a graduated transparent plate. The performing method of the Brinell hardness test is presented in ISO 6506-1 (2014) and ASTM E10-17 (2017) standards. The Brinell hardness is calculated through the following equation:

$$\mathrm{BHN} = \frac{2P}{\pi D\left(D^2 - d^2\right)} \tag{6.4}$$

In this equation, BHN is the Brinell hardness number (kgf/mm^2), P is the load applied (kgf), D is the diameter of the ball indenter (mm) and d is the diameter of the circular effect area (mm) resulting from indentation. The Brinell hardness is sometimes expressed in the MPa unit. In Figure 6.1, a schematic image of the Brinell test device and the necessary parameters to calculate the hardness value can be seen.

As mentioned, in the Brinell method, a large indenter is used and the hardness is obtained from the diameter of the indentation. The large size of Brinell's effect can make it impossible to use this test in small parts or parts that are under critical stress and indentation in them can lead to failure. This test provides acceptable results for determining the surface hardness of rocks that have a heterogeneous surface.

6.2.1.4 *Vickers hardness*

The Vickers test is one of the hardness evaluation methods, which is based on determining the resistance of materials and minerals against a pyramid-shaped indenting object made of diamond. Hardness measurement in this approach was developed by Smith and Sandly (1992) as an alternative to Brinell hardness. In this test, a square-base pyramid is used as a mandrel, whose opposite surfaces meet each other at an angle of 136°. The reason for using this angle is that it is close to the most favorable ratio of the indentation diameter to the ball diameter in the Brinell hardness test. The Vickers hardness test performing method is provided in ISO 6507-1

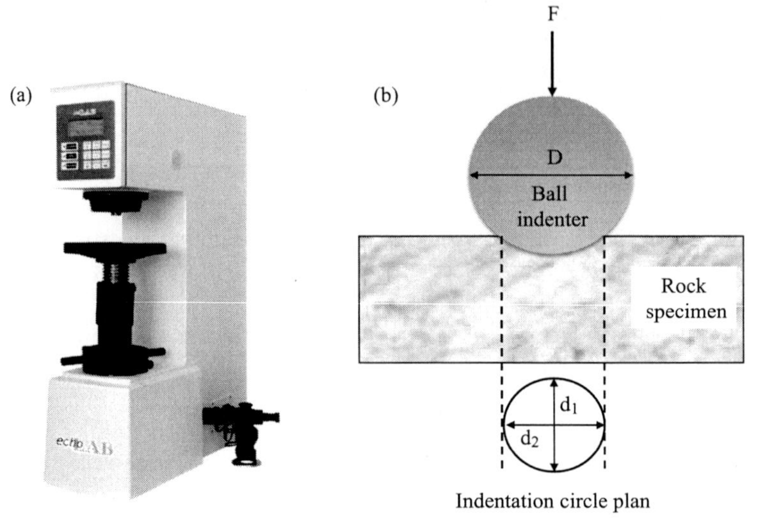

Figure 6.1 (a) Brinell test device and (b) a schematic image of the ball indenter into the rock specimen and the necessary parameters to calculate the hardness.

(2018) and ASTM E92-16 (2016) standards. The surface hardness value using the Vickers test is obtained from the following equation:

$$\mathrm{VH} = \frac{2F\sin\left(\frac{136^\circ}{2}\right)}{d^2} \approx \frac{1.854F}{d^2} \tag{6.5}$$

In this equation, VH is the Vickers hardness number (kgf/mm^2), F is the load applied in kgf and d is the diameter of the square effect area (mm) resulting from indentation. Since rocks are generally heterogeneous materials composed of one or more minerals with different hardness, the Vickers hardness of a rock is a weighted average hardness of its constituent minerals, which can be calculated from the following relationship (Heiniö 1999):

$$\mathrm{VH} = \sum_{i=1}^{n} A_i \times (\mathrm{VH})_i \tag{6.6}$$

In this equation, VH is the Vickers hardness of the rock, A_i is the percentage of each mineral in the rock and $(\mathrm{VH})_i$ is the Vickers hardness of that mineral. The Vickers hardness value of each rock is equal to an average of 3–5 times the microhardness value of its constituent mineral (Fereidooni and Ghasemi 2023). In Figure 6.2, a schematic image of the Vickers test device and the necessary parameters to calculate the hardness value can be seen.

6.2.1.5 Rockwell hardness

The Rockwell hardness test is one of the hardness testing methods for metals, which measures hardness based on the object's resistance to indentation by applying different loads. This method was introduced in 1914 by the Stanley brothers (Rockwell and Rockwell 1914). In this method,

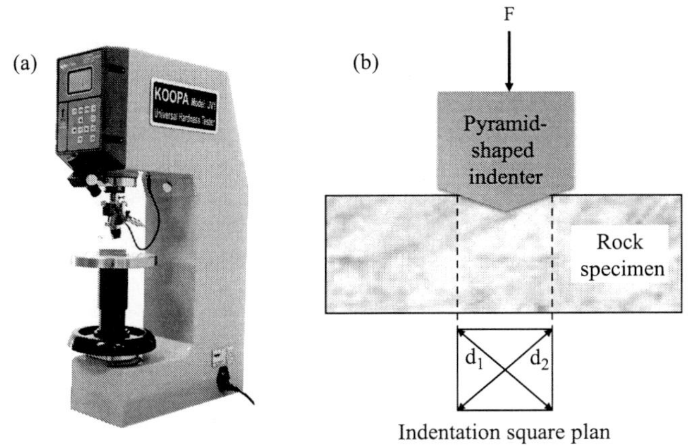

Figure 6.2 (a) Vickers test device and (b) a schematic image of the pyramid-shaped indenter into the rock specimen and the necessary parameters to calculate the hardness.

the tested specimen surface is pressed by the diamond indenter and the initial indentation depth is measured. Then, maintaining the initial force, the main force is applied. After that, the main force is removed, leaving only the initial force, and the final indentation depth is measured. The Rockwell hardness performing test method is presented in ISO 6508-1 (2016) and ASTM E18-15 (2015) standards. It is easy to perform a Rockwell hardness test with acceptable results for minerals and rocks because the initial loading of the sample eliminates errors through elastic recovery. The Rockwell hardness test is similar to the Brinell method, but the difference between the two methods is a smaller load and the shape of the indentation in the Brinell test, where the indentation is smaller and deeper. In the Rockwell test, different types of indenters and forces can be used, and each combination of these two conditions creates a special Rockwell scale. To determine the Rockwell hardness, the following three methods are used:

- Rockwell A: In this method, the amount of force is 60 kg.
- Rockwell B: In this method, the amount of force is 100 kg.
- Rockwell C: In this method, the amount of force is 150 kg.

Table 6.2 shows the different methods for performing the standard Rockwell test. Figure 6.3 also shows the Rockwell test device and types of spherical and conical indenters for performing the test.

6.2.1.6 Knoop hardness

The Knoop hardness test was proposed by Knoop et al. (1939) as an alternative to the Vickers test. The method for determining the Knoop hardness is described in ISO 4545-1 (2017) and ASTM E92-16 (2016) standards. In this test, the surface of the tested specimen is pressed by a pyramidal diamond indenter with a rhombus section. The diamond tip is placed on the surface of the specimen and is subjected to specific forces of 0.001–2 kgf (0.009807–19.61 N) for 10–15 seconds. Then, the average diagonal length of the tip indentation is measured by a microscope (Figure 6.4). It is important to note that the effects of indentation will be different depending on the rock type and the applied load amount. There are three standards, TS EN 14205 (2004),

Table 6.2 Different methods for performing the Rockwell test

Type	*Abbreviation*	*Force (kgf)*	*Indenter*	*Application*
A	HRA	60	120° diamond cone	Thin and hardened steels
B	HRB	100	Steel sphere with a diameter of 1/16 inches	Aluminum alloys, copper alloys and soft steels
C	HRC	150	120° diamond cone	Hard steels, titanium alloys
D	HRD	100	120° diamond cone	Pearlitic malleable iron
E	HRE	100	Steel sphere with a diameter of 1/8 inches	Cast iron, magnesium aluminum alloys
F	HRF	60	Steel sphere with a diameter of 1/16 inches	Annealed copper alloys
G	HRG	150	Steel sphere with a diameter of 1/16 inches	Beryllium-copper, malleable irons
H	HRF	60	Steel sphere with a diameter of 1/8 inches	Aluminum, zinc, lead
K	HRF	150	Steel sphere with a diameter of 1/8 inches	Very soft or thin material
L	HRF	60	Steel sphere with a diameter of 1/16 inches	Very soft or thin material
M	HRF	100	Steel sphere with a diameter of 1/16 inches	Very soft or thin material
P	HRF	150	Steel sphere with a diameter of 1/16 inches	Very soft or thin material
R	HRF	60	Steel sphere with a diameter of 1/2 inches	Very soft or thin material
S	HRF	100	Steel sphere with a diameter of 1/2 inches	Very soft or thin material
V	HRF	150	Steel sphere with a diameter of 1/2 inches	Very soft or thin material

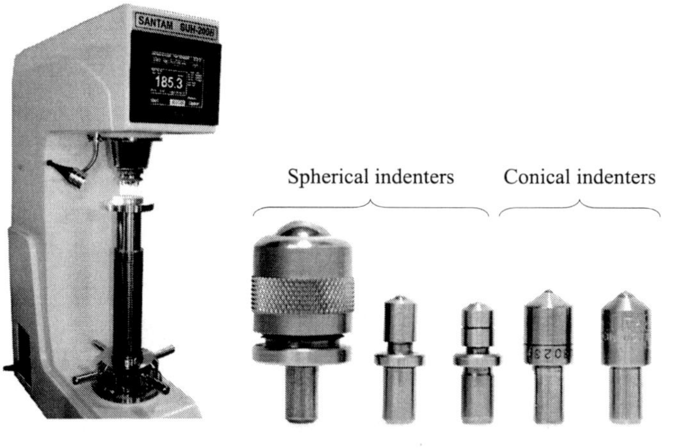

Figure 6.3 Rockwell test device and types of spherical and conical indenters to perform the test.

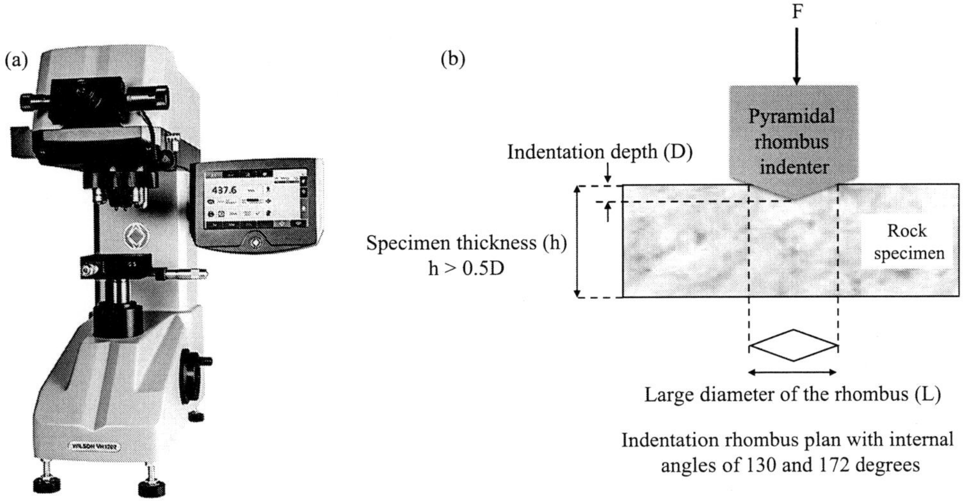

Figure 6.4 (a) Knoop test device and (b) a schematic image of the pyramidal rhombus indenter in the rock specimen and the necessary parameters to calculate the hardness.

BS EN 14205 (2003) and EVS-EN 14205 (2004), for measuring the hardness of natural rocks. At least one polished section of rock approximately 20 mm wide, 30 mm long and 10 mm thick should be prepared based on the British Standard (BS EN 14205 2003) for this test. The hardness value of the surface of the material after the Knoop test is obtained from the following equation:

$$NH = \frac{F}{C_p \times L^2} \tag{6.7}$$

In this equation, NH is the Knoop hardness number (kgf/mm^2), F is the applied load (kgf), C_p is the indentation coefficient equal to 0.070279 and L is the large diameter of the effect resulting from the indentation of the rhomboidal indenter on the surface of the material (mm). Figure 6.5 shows the relationship between the Knoop and Mohs hardnesses.

The important point about the four indentation hardness determination methods, including Brinell, Vickers, Rockwell and Knoop, is the limitation of their use in rock engineering. In other words, these four methods have been developed for application in metallic materials. However, there are still no complete and codified standards other than the Knoop method that can be used in rock mechanics applications. Therefore, to use the mentioned methods in rock mechanics applications, the effect of different parameters and the correlation between them should be carefully investigated. In other words, investigating different parameters of rock materials on a laboratory scale in a wide range of rocks is necessary to provide standards for the application of each method in rock engineering (Ghorbani et al. 2022). Some of the important parameters that should be taken into consideration in performing the hardness measurement methods mentioned in rock materials are:

- Scale effect (thickness, area and volume of rock specimen)
- The rock specimen shape (core, block or irregular geometric shape)

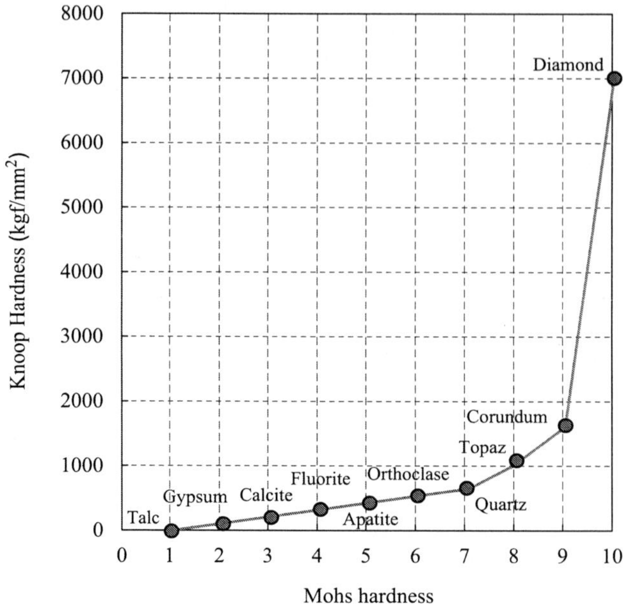

Figure 6.5 Relationship between the Knoop and Mohs hardnesses.

- Choosing the right method of hardness testing on rock specimens
- The number of hardness tests on the rock specimen surface
- Effect of physical, mechanical and textural properties of rock
- Effect of weathering of rock specimen
- The effect of the ambient temperature of the experiment
- The effect of tested specimen surface roughness

6.2.1.7 *Indentation hardness index*

In this method, a rock with a ratio of height to diameter of at least 0.75–1 is placed on the bottom plate of the point load device and loaded. The specimen conical plate has a 60° cone and a spherical tip with a radius of 5 mm. The tip of the cone transfers the charge to the specimen, and a counter reads the resulting penetration area. The indentation hardness value is obtained by dividing the maximum load (kN) by the maximum penetration value (mm) according to the following equation. The proposed classification of rock hardness based on the indentation hardness index (IHI) is shown in Table 6.3. The average of at least three tests on the sample is reported as the indentation hardness value. The indentation hardness index indicates the rock's resistance to elastoplastic deformation. The rock indentation hardness test has undergone many changes and improvements since its initial introduction a few decades ago. More details for this method are provided by Szwedzicki (1998):

$$\mathrm{IHI} = \frac{L}{p} \tag{6.8}$$

In this equation, IHI is the indentation hardness index, L is the applied load (KN) and p is the maximum amount of penetration (mm).

Table 6.3 Classification of rocks based on the indentation hardness index (Szwedzicki 1998)

Indentation hardness index (IHI)	*Description*
> 50	Extremely hard
40–50	Very hard
30–40	Hard
25–30	Approximately hard
20–25	Approximately soft
15–20	Soft
8–15	Very soft
< 8	Extremely soft

6.2.1.8 Cerchar hardness test

The Cerchar hardness test was introduced in the 1970s to test the abrasiveness of coal-bearing rocks in the mining industry by the French Cershar Institute, and it was gradually used in the tunneling industry. Usually, the Cerchar hardness index (CHI) is used to determine the amount of rock abrasiveness to estimate the life of the bit in various mining and tunneling applications. Although the Cerchar test is simple and quick, there are some differences in the test results due to the equipment used, the condition of the rock surface, the skill of the operator and the methods used for measuring the level of abrasiveness. While the initial design of this test was presented by Cerchar (1973), today there are various devices to perform this test. In general, all these devices consist of a clamp that holds the specimen in a position, over which a hardened steel stylus with a 90° conical tip is pulled over its surface under a constant load of 70 N. The steel stylus is drawn on the surface of the rock specimen at a distance of 10 mm. Another device designed by West (1989) is widely used in commercial and laboratory applications.

Cerchar (1986) suggested the use of two to three tests for fine-grained and homogeneous rock samples and five or more tests for rock samples with grain size greater than 1 mm. West (1989) indicated that five tests should be performed on each sample, and most researchers and commercial laboratories follow this recommendation. Finally, the average measurements are reported as the test result. The expansion caused by the wear of the steel stylus tip is measured using a microscope. The diameter of the stylus tip should be measured in two perpendicular directions, and its average value should be determined. By performing the test and determining the average spread of the tips of the styluses, the value of the Cerchar abrasivity index (CAI) is obtained from the following equation:

$$\mathrm{CAI} = 10 \times d \tag{6.9}$$

In this equation, CAI is the Cerchar abrasivity index and d is the diameter of the stylus tip (mm). Based on the Cerchar abrasivity index, the degree of abrasiveness for different rocks can be described according to Table 6.4.

In another method of performing the Cerchar hardness test, a tungsten carbide drill bit with a diameter of 8 mm and a tip angle of 99° is rotated on the rock specimen with a force of 200 N, and the time to drill a hole with a depth of one centimeter assuming a constant rotation speed is considered the CHI (Ghorbani et al. 2022). In Table 6.5, the classification of rocks based on the CHI determined in this way is presented.

Table 6.4 Classification of abrasiveness degree of rocks based on the Cerchar abrasivity index (Cerchar 1973)

Cerchar abrasivity index (CAI)	*Description*
<0.3	Non-abrasive
0.3–0.5	A little abrasive
0.5–1	Moderate abrasive
1–2	Abrasive
2–4	Very abrasive
4–6	Extremely abrasive
>6	Quartz

Table 6.5 Classification of rocks based on the Cerchar hardness index (Sofretu and Schwenzfeier 2005)

Cerchar hardness index (CHI)	*Description*
0–20	Soft
21–40	Approximately hard
41–80	Hard
81–120	Very hard
> 120	Extremely hard

Various factors are effective on the results of the Cerchar hardness test, the most important of which are as follows:

- Rock composing minerals: Increasing the hardness of the composing minerals causes an increase in the Cerchar abrasivity index.
- Size of rock crystals or grains: Increasing the size of crystals and grains of rock increases the Cerchar abrasivity index.
- Roundness of rock grains: The presence of rounded grains in a rock reduces the Cerchar abrasivity index.
- Rock compressive strength: There is a direct and linear relationship between the Cerchar abrasivity index and the rock compressive strength.
- Surface roughness of rock specimen: The Cerchar abrasivity index for the natural fracture surfaces of rock is higher than its value for the sawed surfaces.
- Length of pulling the steel stylus: Increasing the length of pulling the steel stylus on the rock surface increases the Cerchar abrasivity index. However, most of the expansion of the stylus tip occurs at the beginning of the test, and the expansion rate gradually decreases.
- Pulling speed of the steel stylus: Increasing the pulling speed of the steel stylus on the rock surface approximately causes a decrease in the Cerchar abrasivity index.
- Hardness of the device steel stylus: As the hardness of the used steel stylus increases, the Cerchar abrasivity index decreases. The dispersion of the results obtained when using harder styluses is less.
- Rock confining stresses: As the rock confining stresses increase, the Cerchar abrasivity index increases. This increase is more evident in rocks with higher porosity.

6.2.1.9 Cone indentation test

The cone indentation test was designed as a standard in the former National Coal Mining Research and Development Institute of England to test the resistance of rock and coal to the indentation of a tungsten carbide cone with a tip angle of 60°. This method determines the hardness of small pieces of stone with dimensions of 12 × 12 × 6 mm. Considering the displacement between the first and second advance of the indentation cone (M_1 and M_2), the cone indentation hardness is calculated using the following equation (Szlavin 1974):

$$\text{CIH} = \frac{0.635}{\left(M_1 - M_2\right) - 0.635} \tag{6.10}$$

In this equation, CIH is the cone indentation hardness index and M_1 and M_2 are the first and second indentation values of the cone (mm), respectively. One of the limitations of this method is the low indentation depth of the cone in hard rocks. For this reason, Szlavin (1974) modified this method. In the modified version, the indentation force applied was increased to 110 N. This method does not provide good results in coarse-grained rocks.

6.2.1.10 Nanoindentation test

Today, the nanoindentation test has become a common method for determining the mechanical properties of homogeneous and heterogeneous materials. This test includes loading a specimen by applying a specified force by the user and determining the indentation depth and then unloading the specimen. There are mainly two types of indenters in the nanoindentation test, which are Berkovich and corner of cube. The Berkovich indenter is a three-sided pyramid with a surface angle of 63.5° to the vertical indentation axis, and its surface-to-depth ratio is the same as the Vickers test indenter (Berkovich 1951). The corner of cube is also a three-sided pyramid, which is exactly the corner of a square cube. The nanoindentation test is suitable for checking the surface hardness of homogeneous materials such as metals, minerals or mono-mineral rocks.

The nanoindentation test can be performed at a constant loading rate or a constant strain rate. In the case of a constant loading rate, there is a phase of elastic deformation when the nano-indenter starts to compress the rock surface. After that, increasing the load leads to plastic deformation, and according to the geometry of the indentations, a permanent indentation can be observed. The constant strain rate mode depends on the application of rapid high-frequency oscillation to the quasi-static loading signal. Therefore, a harmonic force can be added to the load. The main advantage of this method is that it provides a direct measurement of the dynamic hardness at any point along the loading curve.

6.2.2 Dynamic hardness tests

The dynamic hardness testing methods include a main group called rebound hardness testing methods. In measuring the hardness with dynamic methods, usually, an object falls or hits on the surface of the tested material and the hardness of the surface is determined according to the height of its return, which itself indicates the amount of energy absorbed by the surface of the material. The Shore scleroscope is the most common dynamic hardness testing device, and it measures the hardness according to the rebound height of a ball. Also, the Schmidt rebound hardness, which is used to determine rock surface hardness, rock joint surface strength and concrete hardness, is one of the other types of dynamic hardness measurement methods. The Leeb hardness test is also included in the category of dynamic hardness measurement methods (Fereidooni and Ghasemi 2023).

6.2.2.1 Shore scleroscope

The Shore scleroscope test was first presented in 1907 by Albert F. Shore. The Shore hardness test is accepted as an easy and non-destructive method to measure rock hardness and is widely used in rock mechanics. The Shore test device is provided in two types, C and D. The hardness tester consists of a small diamond-tipped hammer and a graded glass tube. A half-ounce (14.2 g) hammer is dropped freely from a height of 10 inches (250 mm) through a graduated glass tube onto the surface of the test specimen. The glass tube is divided into 140 equal distances. Therefore, the Shore hardness test ranges from zero to 140 on a calibrated scale. The rebound height is read by eye judgment from the graduated lines of the tube and is considered the hardness number of the specimen. To get a reliable result, it is necessary to perform several consecutive tests. For each impact, the location of the testing device on the specimen surface must be moved (changing the location of the impact).

Researchers use standard sample sizes to optimize test results in the laboratory. Misra (1972) reported that rock specimens with a diameter of 25 mm, a surface area of 91.4 cm^2 and a length of 5 cm should be used for the Shore hardness test. According to the ISRM (1978) standard, the minimum surface area of the rock specimen and its minimum thickness should be 10 cm^2 and 1 cm, respectively. Later, Rabia and Brook (1979) determined the minimum specimen volume to be 40 cm^3. Holmgeirsdottir and Thomas (1998) investigated the effectiveness of the D-type Shore scleroscope for testing rock hardness in small volumes. Their results show that not only is type D easier than type C, but it can also be used for small specimens. Altindag and Guney (2006), in an experimental study, suggested the critical volume of specimens tested with the Shore scleroscope is equal to 80 cm^3. Figure 6.6 shows a schematic image of the Shore Scleroscope and its working mechanism.

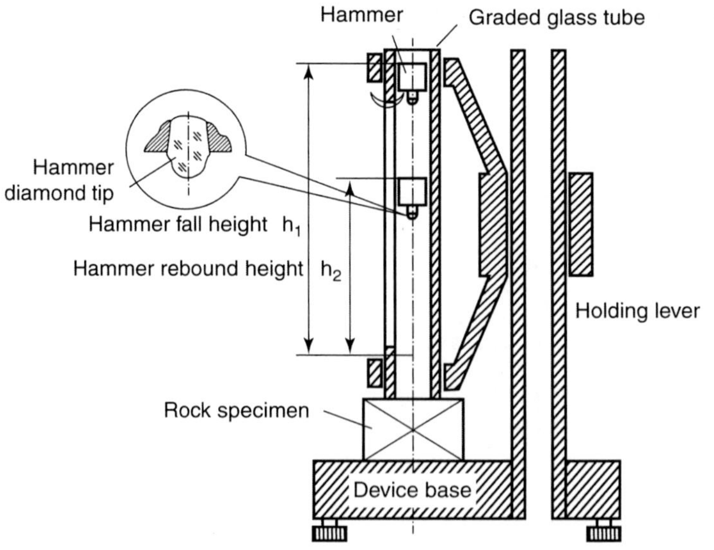

Figure 6.6 A schematic image of the Shore scleroscope device and its working mechanism.

6.2.2.2 *Schmidt rebound hardness*

The use of this device was first performed by a Swiss engineer named Ernst Schmidt (Schmidt 1951), and now it is known as the Schmidt hammer test, the performing method of which is standardized by ISRM (1978) and ASTM-D-5873-14 (2005). Based on Figure 6.7, the Schmidt hammer has a steel rod that is placed in contact with the surface of rock or rock joint surface. Inside the hammer, there is a weight that hits the steel bar with a certain amount of energy, and after hitting, the weight opens, and meanwhile, the index connected to the weight shows the amount of rebound. In fact, the mechanism of the Schmidt hammer is based on hitting a steel rod (anvil) on the surface of the joint and its rebound and reading the corresponding number on the hammer index. The value of the number read in this test is the distance of the rebound of the weight, which depends on the hardness value of the surface that the steel rod of the hammer hits. According to the mechanism of action and the amount of energy released by the Schmidt hammer, there are two types of Schmidt hammers: L-type and N-type. The effective energy of the L-type ones is equal to 0.735 Nm, and it is only one-third of the impact energy of the N-type hammer, with an impact energy of 2.207 Nm (Khanlari and Fereidooni 2011).

The rebound number can be related to the compressive strength of the rock by the calibration curve on the hammer. The calibration curve for compressive strength in this method has a very high sensitivity and is influenced by several factors, including rock type, unit weight, porosity, moisture percentage, degree of weathering, the length of the hammer's placement concerning the surface of the rock during the test and so forth (Fereidooni and Khajevand 2023).

Similar to the Shore hardness, many research efforts have been made to provide a standard specimen size in the Schmidt hardness test. The length of the tested block edge in this test according to ISRM (1981) and ASTM D5873-14 (2005) suggested methods should be 6 and 15 cm, respectively. Also, Aydin (2008) and Demirdag et al. (2009) suggested that the edge dimensions of the rock specimen should be 10 and 11 cm, respectively.

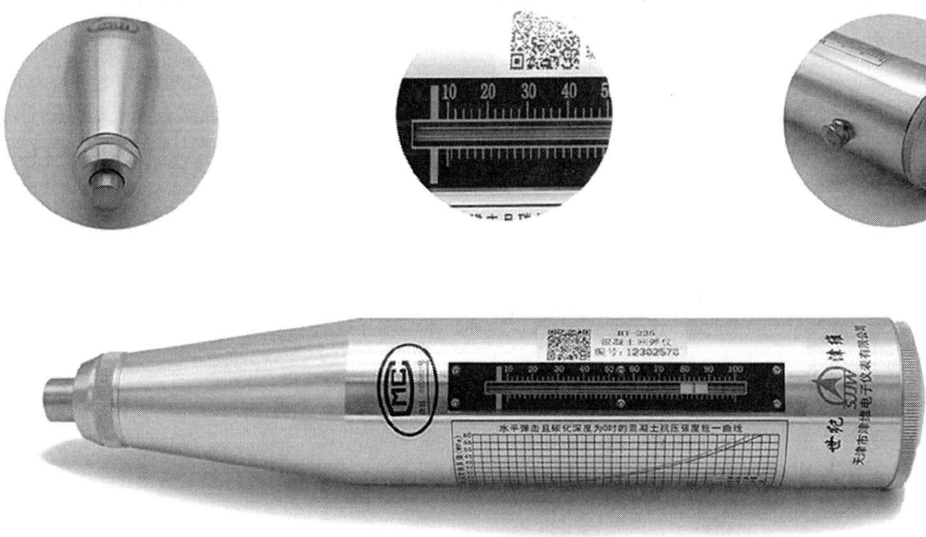

Figure 6.7 An image of the Schmidt hammer and its components.

When using the Schmidt hammer on rock surfaces, at least 20 impacts should be done for each sample, then half of the data with the lowest values are removed and the rest are averaged. Since the application direction of the Schmidt hammer is effective on the rebound number, the results should be corrected based on the hammer application direction according to Table 6.6. For this purpose, the values that have a positive sign in the table are added to the average value of the returned number, and the values with a negative sign are subtracted from the average value. If the hammer is applied horizontally, there is no need to correct the rebound number. Also, if the calculated value is not a multiple of 10, interpolation should be used to calculate the correction value. In Table 6.7, the range of the Schmidt hardness values for some common rocks is presented for comparison.

One of the important applications of the Shore scleroscope and Schmidt hammer is to determine the plasticity of rocks. For this purpose, McFeat-Smith (1977) introduced the plasticity index (PI) using the Shore hardness. In this method, after repeatedly hitting a point on the rock specimen surface, a plastic deformation surface is created on the specimen, and the surface hardness increases in that place. The PI can also be measured using the Schmidt hammer according to the following relationship (Yasar 2020):

$$PI = \frac{Q_{20} - Q_1}{Q_1} \times 100 \tag{6.11}$$

Table 6.6 Schmidt hardness correction for non-horizontal surfaces (quoted from Fereidooni 2023)

Number of rebounds	*Correction value for deviation angle*			
	Upward		*Downward*	
	+90	*+45*	*−45*	*−90*
10	–	–	+2.4	+3.2
20	−5.4	−3.5	+2.5	+3.4
30	−4.7	−3.1	+2.3	+3.1
40	−3.9	−2.6	+2.0	+2.7
50	−3.1	−2.1	+1.6	+2.2
60	−2.3	−1.6	+1.3	+1.7

Table 6.7 Value of the Schmidt rebound hardness of some common rocks

Group	*Rock type*	*Schmidt rebound hardness*
Igneous	Granit	54–69
	Andesite	67
	Basalte	61
Sedimentary	Sandstone	10–37
	Mudstone	27
	Limestone	35–51
Metamorphic	Gniess	49
	Schist	31

In this equation, PI is the plasticity index of the rock specimen, and Q_1 and Q_{20} are 1st and 20th Schmidt rebound hardness, respectively. A larger PI means a higher plastic property of the rock specimen.

6.2.2.3 Leeb hardness

In the Leeb or Equotip hardness test, the impact is transferred from a ball made of diamond or tungsten carbide to the rock specimen surface (Figure 6.8). In this method, the Leeb hardness number is the ratio of the rebound velocity of the ball to its impact velocity, which can be calculated from the following equation in accordance with the ASTM A956-06 (2006) standard:

$$\mathrm{LH} = \frac{v_R}{v_I} \times 1000 \tag{6.12}$$

In this equation, LH is the Leeb hardness, v_R is the ball rebound velocity and v_I is the impact velocity of the ball. Six types of impact devices are used in the Leeb hardness test, which includes D, S, E, DL, D+15, G and C. Since the impact exerts force vertically on the surface of the specimen, the hardness values for other directions must be corrected. According to ASTM A956-06 (2006), the three basic points that should be considered in the Leeb hardness test are as follows:

- The surface of the specimen must be carefully prepared to avoid changes in hardness.
- Magnetic fields affect the results of this test. It is recommended that the residual magnetic field is less than 4G.
- To perform the test correctly, the specimen should not be subjected to vibration.

The impact energy of the Leeb D-type testing device is approximately equivalent to 1/200 and 1/66 of the N-and L-type Schmidt hammers, respectively. Therefore, it causes less damage to the tested surface and provides acceptable results in soft and weak rocks. Also, the Schmidt hammer and Leeb hardness tester are both easy to use in the field and allow many readings to be taken on relatively large surfaces of rock in a short time.

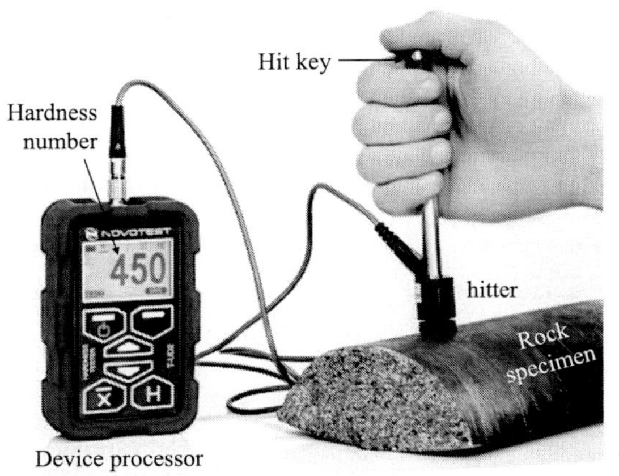

Figure 6.8 An image of the Leeb hardness device and its application mechanism on the rock surface.

6.2.3 Other hardness tests

Apart from the hardness measurement methods discussed above, there are other methods for this purpose. Taber abrasion hardness test, total rock hardness, Protodyakonov scale of hardness and rock impact hardness number (RIHN) are other hardness measurement methods that are less used in rock mechanics.

1. Taber hardness test: To determine the Taber abrasion hardness, each side of two disc-shaped specimens of NX size (thickness 0.25 inches or 6 mm) rotates 400 times under an abrasive wheel that presses the discs with a weight of 250 g. Then, the weight loss of the specimens is measured and their average is calculated. In other words, the Taber abrasion number is the inverse of weight loss. This test is sensitive to factors that affect strength, shearing, crushing and abrasion on a small scale. The Taber hardness is calculated according to the following equation (Tarkoy 1973):

$$H_A = \frac{1}{\text{AWL}} \tag{6.13}$$

   In this equation, H_A is the Taber hardness and AWL is the average weight loss of disc-shaped specimens in g.

2. Total rock hardness: The total rock hardness was presented by Tarkoy (1973) from the combination of the Schmidt rebound and Taber hardness methods. This method is developed to predict rock performance to estimate the excavation rate of the tunnel boring machine (TBM) and is calculated from the following equation:

$$H_T = \text{SRH} \times \sqrt{H_A} \tag{6.14}$$

   In this equation, SRH is the Schmidt rebound hardness, and H_A is the Taber abrasion hardness.
3. Protodyakonov hardness scale: The Protodyakonov hardness scale has been used less than other methods by rock engineering researchers. This scale indicates the relative value of rock resistance against fracture ($f=R/10$) (where f is the relative hardness and R is the compressive strength of rock). According to the Protodyakonov hardness scale, the hardness of rock is divided into five degrees according to Table 6.8.
4. Rock impact hardness number (RIHN): The RIHN was proposed by Brook and Misra (1970) as a modified method of the Protodyakonov index. In this method, first, rock cores with a diameter of 25 mm and a length of 50 mm are prepared. Then, the cores are air-dried and weighed. Each core specimen is placed horizontally in the mortar along the axis of the cylinder and is broken by the impact of a steel weight of 2.4 kg. Then, the pieces of fine grains produced with a size smaller than 500 μm are weighed and the weight percentage of the fine grains is calculated relative to the initial weight of the core. This method is repeated for a

Table 6.8 Degrees of rock hardness based on the Protodyakonov hardness scale (Zhang et al. 2017)

Hardness degree	*Very hard*	*Hard*	*Semi-hard*	*Soft*	*Very soft*
f value	>20	16–20	8–16	2–8	<2

number of different impacts and continues until the percentage of fine particles produced in each test reaches more than 30%. In the next step, a graph of the percentage of small grains (y-axis) against the number of impacts (x-axis) is prepared. Finally, the RIHN is defined as the number of impacts to produce 25% of fine-grained fragments. Brook and Misra (1970) suggested that only one measurement should be taken for each number of impacts and that four to five cylindrical rock specimens are sufficient to determine the RIHN.

6.3 Mandrel indentation hardness test

The mandrel indentation test (MIT) was proposed by Fereidooni (2022a). Most of the hardness testing methods were first presented for metals and non-metallic materials. In fact, in most indentation tests, the size of the indentation created is small, and it seems that using a larger indenter for materials with a heterogeneous surface, such as natural rocks and building stones, can provide more accurate results. Therefore, in this method, a larger mandrel that creates a larger conical indentation is used to perform the indentation hardness test on the rock surface. This test is named as Mandrel indentation test. To perform the test and calculate the mandrel indentation hardness (MIH), a metal frame made of Mo40 hard steel was designed and manufactured in a turning workshop. This frame includes two upper and lower jaws, an indentation mandrel, a pressure gauge, a locking screw, and sliding bars. The frame is placed in an electrical jack that provides pressure (Figure 6.9). The mandrel used in this device is made of very hard steel. The cylindrical shape of the mandrel is 27 mm long, 3 mm in diameter, and has a conical head with an apex angle of 30° (θ=30°), which is fixed to the upper jaw of the frame. A rock specimen is placed within the frame and between the lower jaw and mandrel. An arbitrary axial force (20 kg is a good value) is applied by an electric jack and the mandrel is penetrated into the rock specimen. The penetration depth of the mandrel corresponding to a given load is determined in millimeters by a strain gauge. The loading rate is controllable by the jack, so the test time is adjusted by the experimenter.

The procedure for calculating the MIH is similar to other previous indentation tests. In fact, for calculating surface hardness, the applied force must be divided into impression areas. For calculating MIH, based on Figure 6.10, first, the impression area, which is the lateral area of OBB′ cone (A), must be calculated, and the force F is divided into this area. To better understand of the procedure, it can be written:

$$\mathrm{OB} = \frac{\mathrm{BH}}{\sin\left(\frac{\theta}{2}\right)} = \frac{d}{2\sin\left(\frac{\theta}{2}\right)}$$

The value of d (diameter of impression area) can be calculated as follows:

$$\tan\left(\frac{\theta}{2}\right) = \frac{\mathrm{BH}}{\mathrm{OH}} = \frac{d}{2I} \rightarrow d = 2I.\tan\left(\frac{\theta}{2}\right)$$

The value of A (lateral area of OBB′ cone) can be calculated as below:

$$A = \frac{\pi d}{2} \times \mathrm{OB} = \frac{\pi d^2}{4\sin\left(\frac{\theta}{2}\right)} = \frac{\pi I^2.\tan^2\left(\frac{\theta}{2}\right)}{\sin\left(\frac{\theta}{2}\right)}$$

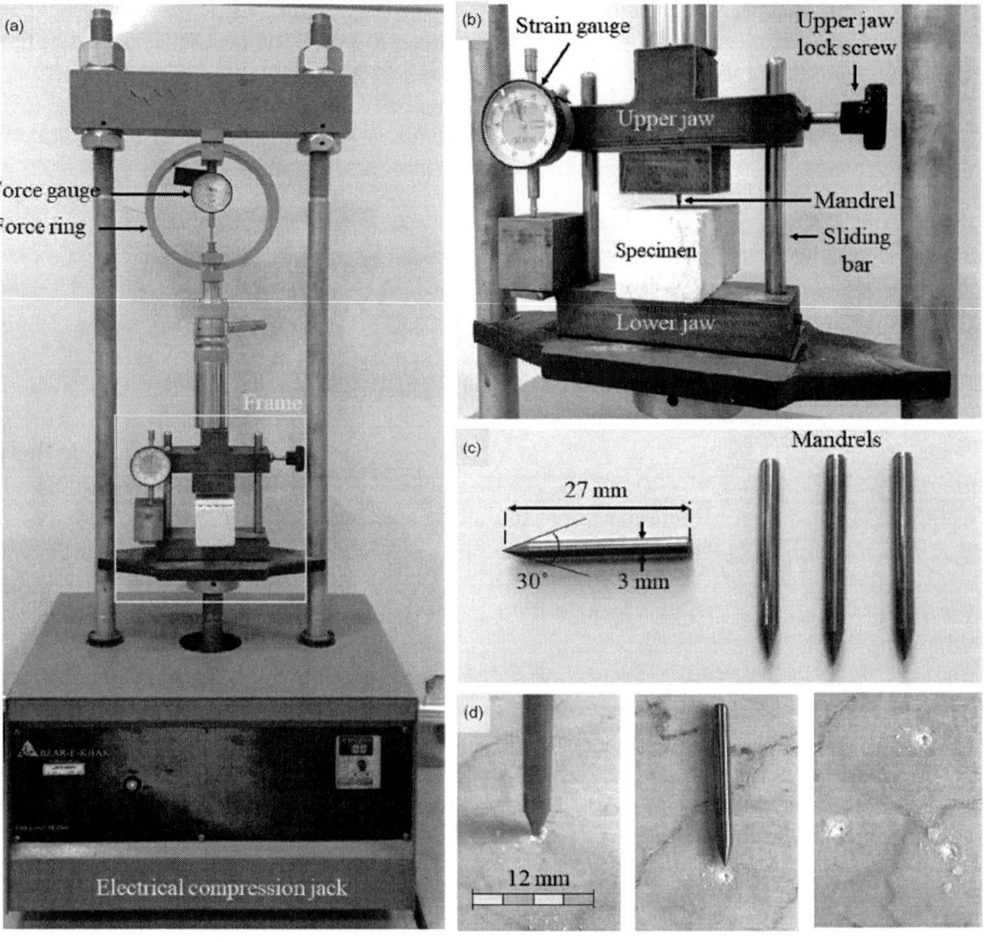

Figure 6.9 Apparatus and test procedures of the mandrel indentation test: (a) the electrical compression jack, (b) the built frame and specimen under the mandrel, (c) the applied mandrels and (d) the indentation impressions on a tested limestone specimen.

Then, the value of MIH is equal to:

$$\mathrm{MIH} = \frac{F}{A} = \frac{F.\sin\left(\frac{\theta}{2}\right)}{\pi I^2.\tan^2\left(\frac{\theta}{2}\right)} \tag{6.15}$$

For $\theta = 30°$, the above equation is simplified as follows:

$$\mathrm{MIH} = \frac{1.147F}{I^2} \tag{6.16}$$

In these equations, F is the applied force in kgf and I is the indentation depth of the mandrel in mm. The values of MIH for the studied building stones were calculated using $F = 20\,\mathrm{kg}$.

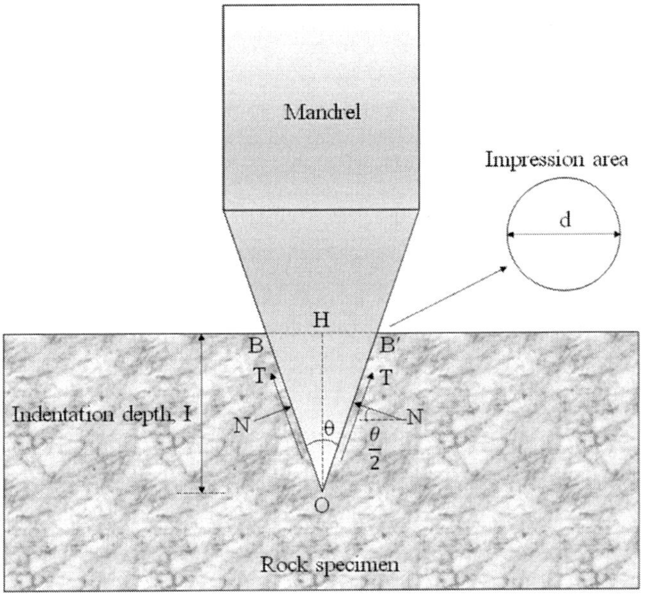

Figure 6.10 Indentation of the mandrel into rock specimen for calculating the mandrel indentation hardness.

6.4 Tests to determine the abrasiveness of aggregates

Aggregates are a collection of small rock pieces, whose diameter varies between 0.075 mm (i.e., sieve no. 200) and about 75 mm (i.e., sieve 3-inch). In engineering works, aggregate is used for the preparation of concrete, preparation of bituminous concrete or asphalt, road and railway infrastructure, dam construction (embankment dams in the form of filters and drainages and sometimes dam shells), airports, construction of well filters, behind of retaining walls as draining and filling materials and mosaic and tabulation. Determining the abrasiveness of aggregates against destructive factors is an important issue. Abrasiveness refers to the response of rocks to physical destructive factors, which shows the amount of rock erosion against erosive factors. In other words, abrasiveness can be defined as the continuous and unwanted loss and removal of materials from the surface of a solid body, as a result of mechanical actions, such as contact and relative movement between two bodies. The interaction of surfaces in contact during relative movement is a general issue that is present in most engineering designs. Abrasiveness characteristics of rocks and aggregates are considered as their behavioral characteristics. Abrasiveness is a scale for measuring the quality of rock materials and aggregates and also a measure for predicting the reaction of rock against drilling and quarrying operations. Evaluation of the qualitative and quantitative properties of aggregate materials can be done by various tests such as the soundness test (sodium sulfate and magnesium sulfate), the Los Angeles abrasion test, the thaw and freezing test, the slake-durability test and the Deval test. Examining the physical and mechanical properties of aggregates determines their use for various purposes. The materials used in roads and flooring should be stable against the abrasion and impact caused by the weight of the upper layers as well as the pressures. Abrasion-based tests are methods that can show the stability of materials against abrasion and failure caused by impact and provide an acceptable result to obtain the quality of materials for use in various construction projects. The abrasiveness of rocks depends on the type and amount of minerals, number of microcracks, rock texture, degree of rock weathering and the adhesion between its grains. Also, the mineralogical composition and physical and mechanical properties of the rocks affect their abrasiveness and strength properties. Grain size is considered as one of the most significant geological factors to evaluate rock abrasiveness.

6.4.1 Deval abrasion test

The Deval test was invented in 1896 to determine the quality of aggregates in road and railway engineering, and then its performing method was presented in the standards of ASTM D6928-10 (2010), BS EN 1097-1 (2010), AASHTO T327-06 (2012), and guidelines presented by the Ministry of Transportation of Ontario, Canada. In this test, five kilograms of aggregate samples are poured inside the rotating cylinder of the test device, which has an internal diameter of 20 cm and an internal length of 34 cm, and it is circulated for 5 hours at a speed of 2000 rotations per hour (10,000 rotations in total). The cylinder axis of the machine has an angle of 30° to the horizon level (Figure 6.11). The Deval test can also be performed in the presence of water, which results in determining the saturation Deval coefficient. After performing the test, the abrasion percentage is calculated through the following equation:

$$W_D = \frac{A-B}{A} \times 100 \tag{6.17}$$

In this equation, W_D is the abrasion percentage, A is the dry weight of the sample remaining on the sieve d (5 kg) before the test, B is the dry weight of the sample after the test and remaining on the sieve d and d is the desired sieve diameter. The sieve diameter d is 1.6 mm according to the French railway regulations and 2.36 mm according to the British railway regulations. After calculating the abrasion percentage, the value of the Deval coefficient (Q) can be calculated through the following equation:

$$Q = \frac{40}{W_D} \tag{6.18}$$

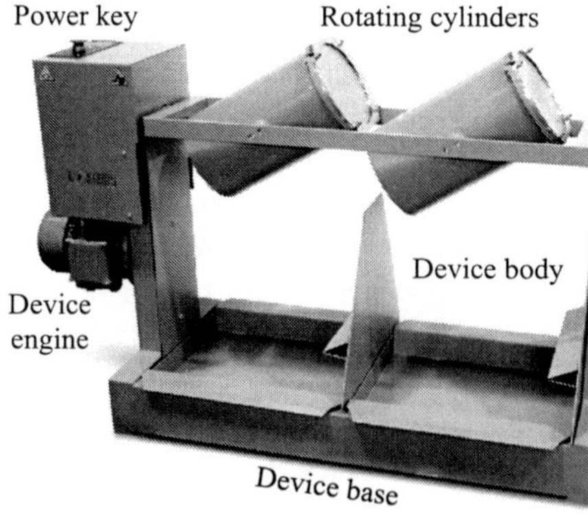

Figure 6.11 Deval test device and its components.

6.4.2 *Los Angeles abrasion test*

The Los Angeles abrasion test was invented in 1926, and its working basis is similar to the Deval test. To assess the quality of aggregate materials, the Los Angeles abrasion test is performed according to ASTM C535-16 (2016) and AASHTO T96 (1996) standards. The Los Angeles test steel cylinder has a diameter of 71.1 cm and a length of 50.8 cm, and its rotation axis is horizontal. Inside the cylinder of the device, there is a 25 mm thick internal thin blade. In this test, 5 kg of dry aggregate with a particle size between 38 mm (sieve 1.5-inch) and 19 mm (sieve no. 8) and passing through specific sieves along with 6–12 steel spheres weighing 416.67 g is poured inside the cylinder and it rotates for 15 minutes at a speed of about 33 rotations per minute (500 rotations in total). Then, the Los Angeles abrasion coefficient is calculated through the following equation:

$$W_{\mathrm{LA}} = \frac{A - B}{A} \times 100 \tag{6.19}$$

In this equation, W_{LA} is the Los Angeles abrasion coefficient, A is the initial weight of the sample in a dry state, and remaining on a certain sieve, and B is the dry weight of the sample after the test and remaining on a certain sieve. The Los Angeles test is performed in four methods of A, B, C and D, and the size of the sieves and the number of steel spheres used in each method are different (Fereidooni et al. 2018). Figure 6.12 shows an image of the Los Angeles test device and the aggregates remaining on certain sieves before the test. Sieves and the amount of aggregate required in each method are presented in Table 6.9, and the number of steel spheres used in each method is also presented in Table 6.10. The Los Angeles test has the following advantages compared to the Deval test:

- Its energy is higher than the energy of the Deval test, and it reveals the weakness points of the samples better.
- The test time is shorter.
- It greatly reduces the role of the experimenter.
- Test and implementation results have a better agreement with each other in different projects.
- It is equally suitable for different materials, including hard rocks, crushed rocks and gravel.

Figure 6.12 Image of (a) Los Angeles test device and (b) samples before testing and remaining on each sieve.

Table 6.9 Aggregate grading required to perform the Los Angeles test

Sieves used		*Required mass of aggregate in each method (g)*			
Passing through the sieve	*Remaining on the sieve*	*A*	*B*	*C*	*D*
$1\frac{1}{2}$ inch	1 inch	1250 ± 25	–	–	–
1 inch	$\frac{3}{4}$ inch	1250 ± 25	–	–	–
$\frac{3}{4}$ inch	$\frac{1}{2}$ inch	1250 ± 10	2500 ± 10	–	–
$\frac{1}{2}$ inch	$\frac{3}{8}$ inch	1250 ± 10	2500 ± 10	–	–
$\frac{3}{8}$ inch	$\frac{1}{4}$ inch	–	–	2500 ± 10	–
$\frac{1}{4}$ inch	No. 4	–	–	2500 ± 10	–
No. 4	No. 8	–	–	–	5000 ± 10
Total mass		5000 ± 10	5000 ± 10	5000 ± 10	5000 ± 10

Table 6.10 The number of steel spheres used in the Los Angeles test

Test method	*Number of steel spheres used*	*Mass of steel spheres used (g)*
A	12	500 ± 25
B	11	4583 ± 25
C	8	3333 ± 25
D	6	2500 ± 25

References

AASHTO T327, 2012. *Standard Method of Test for Resistance of Coarse Aggregate to Degradation by Abrasion in the Micro-Deval Apparatus*. American Association of State Highway and Transportation Officials, Washington, DC.

AASHTO T96, 1996. *Resistance to Degradation of Small-Size Coarse Aggregate by Abrasion and Impact in the Los Angeles Machine*. American Association of State Highway and Transportation Officials, Washington, DC.

Altindag, R., Guney, A., 2006. ISRM suggested method for determining the Shore hardness value for rock. *International Journal of Rock Mechanics and Mining Sciences*, 43,19–22.

ASTM A956-06, 2006. *Standard Test Method for Leeb Hardness Testing of Steel proDucts*. ASTM International, West Conshohocken, PA.

ASTM C535-16, 2016. *Standard Test Method for Resistance to Degradation of Large-Size Coarse Aggregate by Abrasion and Impact in the Los Angeles Machine*. ASTM International, West Conshohocken, PA.

ASTM D5873-14, 2005. *Standard Test Method for Determination of Rock Hardness by Rebound Hammer Method*. ASTM International, West Conshohocken, PA.

ASTM D6928-10, 2010. *Standard Test Method for Resistance of Coarse Aggregate to Degradation by Abrasion in the Micro-Deval Apparatus*. ASTM International, West Conshohocken, PA.

ASTM E10-17, 2017. *Standard Test Methods for Brinell Hardness of Metallic Materials*. ASTM International, West Conshohocken, PA.

ASTM E18-15, 2015. *Standard Test Methods for Rockwell Hardness of Metallic Materials*. ASTM International, West Conshohocken, PA.

ASTM E92-16, 2016. *Standard Test Methods for Vickers Hardness and Knoop Hardness of Metallic Materials*. ASTM International, West Conshohocken, PA.

Aydin, A., 2008. ISRM suggested method for determination of the Schmidt hammer rebound hardness: revised version. In: Ulusay, R. (eds) *The ISRM suggested methods for rock characterization, testing and monitoring: 2007–2014* (25–33). Springer, Chem.

Berkovich, E.S., 1951. Three faceted diamond pyramids for micro-hardness testing. *India's Diamond Industry Review*, 11, 129.

Brinell, J., 1900, Sätt att bestämma kroppars hårdhet jämte några tillämpningar of detsamma. *TeknTidskr*, 30, 69–87.

Brook, N., Misra, B., 1970. A critical analysis of the stamp mill method of determining Protodyakonov rock strength and the development of a method of determining a Rock Impact Hardness Number. *The 12th US Symposium on Rock Mechanics (USRMS)*, American Rock Mechanics Association, St. Louis, MO, pp. 151–166.

BS EN 1097-1, 2010. *Tests for Mechanical and Physical Properties of Aggregates*. British Standards Institute, London.

BS EN 14205, 2003. *Natural Stone Test Methods – Determination of Knoop Hardness*. British Standards Institute, London.

Cerchar, 1986. Centre d'Etudes et Recherches de Charbonnages de France. The CERCHAR abrasiveness index, Verneuil 12S.

Cerchar, A. 1973. Cerchar tests for the measurement of hardness and abrasivity of rocks. *French Coal India*, 73, 1–10.

Demirdag, S., Yavuz, H., Altindag, R., 2009. The effect of sample size on Schmidt rebound hardness value of rocks. *International Journal of Rock Mechanics and Mining Sciences*, 46, 725–730.

Diamantis, K., Fereidooni, D., 2023. Assessing the strength and deformation properties of serpentinite rocks in dry and saturated conditions. *Iranian Journal of Science*, 47(4), 1169–1178.

EVS-EN 14205, 2004. *Natural Stone Test Methods – Determination of Knoop Hardness*. European Standard, Plzen Czech Republic.

Fereidooni, D., Ghasemi, F., 2023. Development of a mandrel indentation test to evaluate surface hardness of dimensional building stones and comparison to Vickers hardness. *International Journal of Rock Mechanics and Mining Sciences*, 170, 05491.

Fereidooni, D., Khajevand, R., 2018. Determining the geotechnical characteristics of some sedimentary rocks from Iran with an emphasis on the correlations between physical, index, and mechanical properties. *Geotechnical Testing Journal*, 41(3), 555–573.

Fereidooni, D., 2016. Determination of the geotechnical characteristics of hornfelsic rocks with a particular emphasis on the correlation between physical and mechanical properties. *Rock Mechanics and Rock Engineering*, 49, 2595–2608.

Fereidooni, D., 2022a. *Design and Build of Mandrel Indentation Test Apparatus for Evaluating Surface Hardness of Building Stones*. Report of Applied and Construction Research Project, Damghan University, Damghan, 50 pp.

Fereidooni, D., 2022b. Importance of the mineralogical and textural characteristics in the mechanical properties of rocks. *Arabian Journal of Geosciences*, 15(7), 637.

Fereidooni, D., 2023. *Engineering Geological*. Damghan University Press, Damghan, 552 pp.

Fereidooni, D., Hashemi, M., Sabouri, V., 2018. Assessing the correlation between slake-durability index and Los Angeles abbreviation for limestone aggregates. *International Conference on Civil Engineering, Architecture and Urban Management in Iran*, University of Tehran, Tehran, Iran.

Fereidooni, D., Karimi, Z., Ghasemi, F., 2024. Non-destructive test-based assessment of uniaxial compressive strength and elasticity modulus of intact carbonate rocks using stacking ensemble models. *PLoS One*, 19(6), 1–29.

Fereidooni, D., Khanlari, G.R., Heidari, M., Sepahigero, A.A., Kolahi-Azar, A.P., 2016. Assessment of inherent anisotropy and confining pressure influences on mechanical behavior of anisotropic foliated rocks under triaxial compression. *Rock Mechanics and Rock Engineering*, 49, 2155–2163.

Fereidooni, D., Khajevand, R., 2023. Evaluating the engineering properties of travertines and their correlations - a case study: Baba-Hafez quarry of Damghan. *The 25th Conference of Geological Society of Iran*, Shahrood University of Technology, Shahrood, Iran.

Fowell, R.J., Abu Bakar, M.Z., 2007. A review of the Cerchar and LCPC rock abrasivity measurement methods. *Proceedings of the 11th Congress of the International Society for Rock Mechanics*, International Society for Rock Mechanics and Rock Engineering, Lisbon, Portugal, pp. 155–160.

Ghasemi, F., Fereidooni, D., 2019. Assessing the block and cylindrical punch shear strengths of some building stones used in Iran. *6th National Conference on Applied Research in Civil Engineering, Architecture and Urban Management*, Khajeh Nasir Toosi University of Technology, Tehran, Iran.

Ghorbani, S., Hoseinie, S.H., Ghasemi, E., Sherizadeh, T., 2022. A review on rock hardness testing methods and their applications in rock engineering. *Arabian Journal of Geosciences*, 15(11), 1–31.

Heiniö, M., 1999. *Rock Excavation Handbook*. Sandvik Tamrock Corp, Stockholm.

Holmgeirsdottir, T.H., Thomas, P.R., 1998. Use of the D-762 shore hardness scleroscope for testing small rock volumes. *International Journal of Rock Mechanics and Mining Sciences*, 35, 85–92.

ISO 4545-1, 2017. *Metallic Materials–Knoop Hardness Test–Part 1: Test Method*. ISO, Geneva.

ISO 6506-1, 2014. *Metallic Materials–Brinell Hardness Test–Part 1: Test Method*. ISO, Geneva.

ISO 6507-1, 2018. *Metallic Materials–Vickers Hardness Test—Part 1: Test Method*. ISO, Geneva.

ISO 6508-1, 2016. *Metallic Materials–Rockwell Hardness Test–Part 1: Test Method*. ISO, Geneva.

ISRM Congress, 2007. *International Society for Rock Mechanics and Rock Engineering*, ISRM, Portugal, pp. 1227–1233.

ISRM, 1978. Suggested methods for determining hardness and abrasiveness of rocks. *International Journal of Rock Mechanics and Mining Sciences & Geomechanics Abstracts*, 15, 89–97.

ISRM, 1981. *Rock Characterisation, Testing and Monitoring*. Pergamon Press, Oxford.

Khajevand, R., Fereidooni, D., 2018. Assessing the empirical correlations between engineering properties and P wave velocity of some sedimentary rock samples from Damghan, northern Iran. *Arabian Journal of Geosciences*, 11, 1–17.

Khajevand, R., Fereidooni, D., 2019. Utilization of the point load and block punch strengths to predict the mechanical properties of several rock samples using regression analysis methods. *Innovative Infrastructure Solutions*, 4, 1–14.

Khanlari, G.R., Fereidooni, D., 2011. Introduce of new equations for calculation of joint compressive strength (JCS) and elasticity modulus in hard rocks, using Schmidt Hammer results. *Journal of Engineering Geology*, 5(1), 1077–1088.

Khanlari, G.R., Heidari, M., Sepahigero, A.A., Fereidooni, D., 2014a. Determination of geotechnical properties of anisotropic rocks using some index tests. *Geotechnical Testing Journal*, 37(2), 242–254.

Khanlari, G.R., Heidari, M., Sepahigero, A.A., Fereidooni, D., 2014b. Quantification of strength anisotropy of metamorphic rocks of the Hamedan province, Iran, as determined from cylindrical punch, point load and Brazilian tests. *Engineering Geology*, 169, 80–90.

Knoop, F., Peters, C.G., Emerson, W.B., 1939. A sensitive pyramidal-diamond tool for indentation measurements. *Journal of Research of the National Bureau of Standards*, 23, 39.

McFeat-Smith, I., 1977. Rock property testing for the assessment of tunnelling machine performance. *Tunnelling and Underground Space Technology*, 9, 29–33.

Misra, B., 1972. Correlation of rock properties with machine performance. PhD Thesis, University of Leeds.

Mucha, K., 2023. Application of rock abrasiveness and rock abrasivity test methods–A review. *Sustainability*, 15, 11243.

Rabia, H., Brook, N., 1979. The shore hardness of rock. *International Journal of Rock Mechanics and Mining Sciences & Geomechanics Abstracts*, 16, 335–336.

Rockwell, H., Rockwell, S., 1914, Hardness tester. USA Patent 129417115.

Schmidt, E., 1951, A non-destructive concrete tester. *Concrete*, 59, 34–35.

Smith, R.L., Sandly, G.E., 1992. An accurate method of determining the hardness of metals, with particular reference to those of a high degree of hardness. *Proceedings of the Institution of Mechanical Engineer*, 102, 623–641.

Sofretu, F.M., Schwenzfeier, G., 2005. *Mechanizated Excavation*. AFTES Working Group, No. 4, Switzerland.

Szlavin, J., 1974. Relationships between some physical properties of rock determined by laboratory tests. *International Journal of Rock Mechanics and Mining Sciences & Geomechanics Abstracts*, 11, 57–66.

Szwedzicki, T., 1998. Indentation hardness testing of rock. *International Journal of Rock Mechanics and Mining Sciences*, 35, 825–829.

Tarkoy, P.J., 1973. A study of rock properties and tunnel boring machine advance rates in two mica schist formations. *Proceedings of 15th Symposium on Rock Mechanics*, Custer State Park, South Dakota, ASCE, New York.

Thuro, K., Plinninger, R.J., 2003. Hard rock tunnel boring, cutting, drilling and blasting: rock parameters for excavatability. *10th ISRM Congress*, OnePetro, Sandton, South Africa.

TS EN 14205, 2004. *Natural Stone Test Methods—Determination of Knoop Hardness*. Turkish Standards Institute, Ankara.

West, G., 1989. Rock abrasiveness testing for tunnelling. *International Journal of Rock Mechanics and Mining Sciences & Geomechanics Abstracts*, 2, 151–160.

Yasar, S., 2020. Predictive plots for conical pick performance using mechanical and elastoplastic properties of rocks. *Journal of Rock Mechanics and Geotechnical Engineering*, 12, 1027–1035.

Yilmaz, N.G., 2011. Abrasivity assessment of granitic building stones in relation to diamond tool wear rate using mineralogy-based rock hardness indexes. *Rock Mechanics and Rock Engineering*, 44, 725.

Zhang, W., Sun, Q., Zhu, S., Wang, B., 2017. Experimental study on mechanical and porous characteristics of limestone affected by high temperature. *Applied Thermal Engineering*, 110, 356–362.

Chapter 7

Seepage and permeability

7.1 Introduction

An important physical property of intact rock, which discusses about the movement of fluids through it, is called permeability. When rock permeability is low, high pore fluid pressures accompanying mechanical and thermal forces are usually valuable. Despite the potential importance of this issue, measuring and interpreting the permeability characteristics of low-permeability intact rock seems a little far-fetched. Because the permeability of the rock mass, which is controlled by its discontinuities, is much more important (Fereidooni and Khajevand 2019). Determining the permeability of intact rock can be done on different scales, from the scale of the earth's crust, which is about 0.5–5 km, to the borehole scale, that is, from 30 to 300 m, and the laboratory scale, which is from 5 to 15 cm. The amount of permeability of large-scale rock in geological environments is influenced by factors such as the frequency of discontinuities, joints, cracks and other geological structures (Fereidooni 2018). Due to the presence of discontinuities in rock mass, the permeability of the rock mass is controlled not only by its composing intact rock, but also by its discontinuities that separate the intact rock blocks from each other. However, on a laboratory scale, measuring the permeability of intact rock requires a correct understanding of the subject mechanism and knowledge of environmental and laboratory factors. The permeability of intact rock is a measure of its capacity to transport fluid. Permeability coefficient or hydraulic conductivity is defined as the discharge rate per unit area under a unit hydraulic gradient. This characteristic depends on the properties of the porous medium (intrinsic factors of rock), fluid characteristics, including its viscosity and density, as well as environmental and laboratory factors. In this point of view, the permeability coefficient of intact rock is different for various fluids depending on their density and viscosity. The permeability of intact rock is usually known as primary permeability and is controlled by rock porosity, which is itself related to factors such as rock type, geological history and in-situ stress conditions. In theory, Darcy's law is used to determine the permeability coefficient of intact rock. Based on the law, the amount of flow through the rock cross-section can be obtained from the following equation (Zhang 2013):

$$Q = K.A.i \tag{7.1}$$

In this equation, K is the permeability coefficient, A is the cross-sectional area of the flow passage and i is the hydraulic gradient which is equal to $i = h/L$ (where h is the hydraulic head and

DOI: 10.1201/9781003672920-7

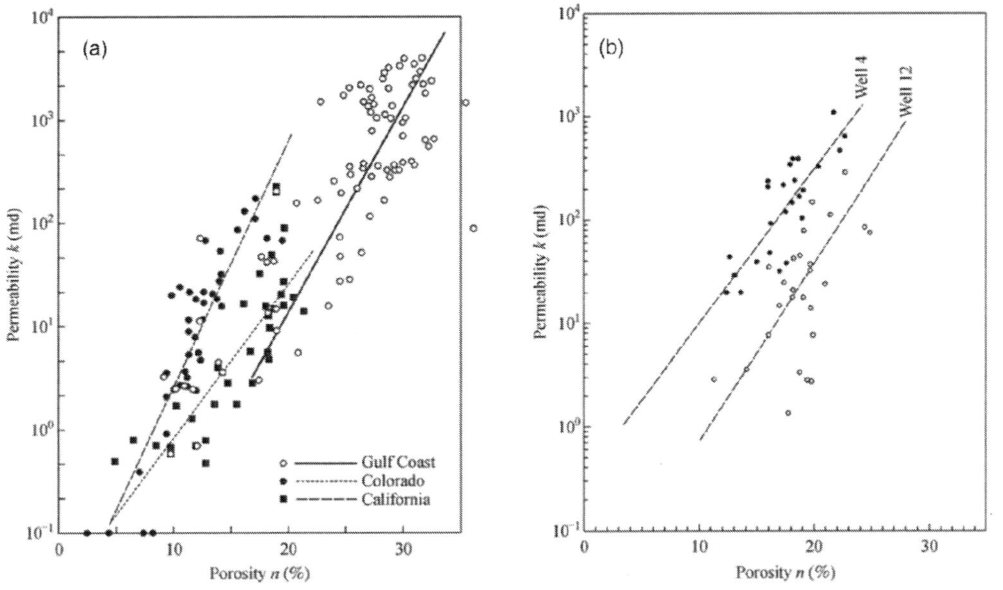

Figure 7.1 Correlations between intrinsic permeability and porosity for (a) three different sandstone samples and (b) Totliegent sandstone (Schön 1996).

L is the length of the flow passage). The permeability coefficient of rock for different fluids, depending on their density and viscosity, varies as follows:

$$K = k \times \frac{\rho.g}{\mu} = k \times \frac{g}{\nu} \tag{7.2}$$

In this equation, k is the intrinsic or specific permeability of rock, which has a square unit of length. ρ, μ and ν are the density, viscosity and dynamic viscosity of the fluid, respectively, and g is the earth's gravitational acceleration equal to $9.81\,\text{m/s}^2$. The intrinsic permeability is independent of the pore fluid properties of rock. Water is often considered as the pore fluid of rock. The intrinsic permeability of intact rock is usually known as primary permeability and, as mentioned, is controlled by the porosity of the rock. Figure 7.1 shows two examples of the relationship between intrinsic permeability (k) and porosity (n) for different sandstones. As can be seen, the relationships between the parameters follow a direct linear relation.

The permeability coefficient of intact rock also changes with the size of the rock grains and increases with increasing larger grain amounts. Figure 7.2 shows the variation of intrinsic permeability (k) versus rock grain size (d). The change process of these parameters can be shown with the following relationship (Schön 1996):

$$\log k = 2.221 \log d - 2.101 \tag{7.3}$$

In this equation, k is in milli-Darcy (10^{-15} square meters) and d is in micrometers. Theoretical and experimental data show that at a given porosity, the pore size distribution of rocks is primarily controlled by their grain size (Yang and Aplin 2010). At a given porosity, the pore size and thus the permeability decrease with decreasing grain size. Using the clay-sized grain amounts

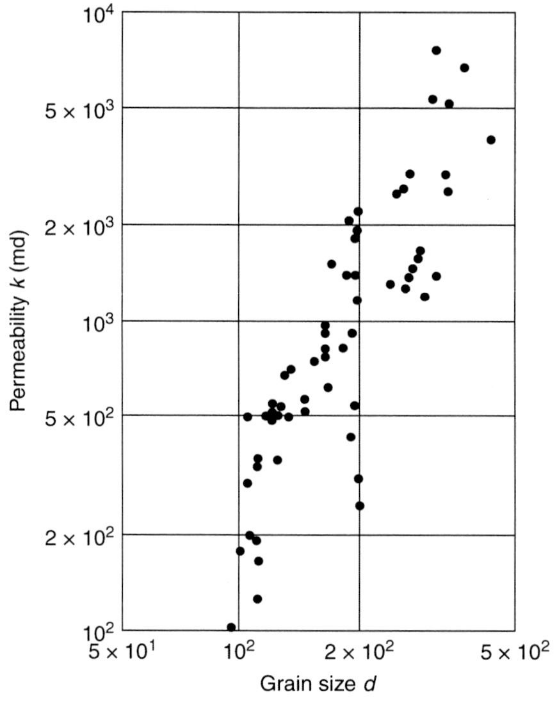

Figure 7.2 Correlation between the intrinsic permeability and grain size for Bentheim sandstone, Germany (Schön 1996).

in rock, Yang and Aplin (2010) presented the following relationship that relates the intrinsic permeability (k), void ratio (e) and clay content (CF) in mudstones:

$$\begin{aligned}&\log k = 69.59 - 26.79\mathrm{CF} + 44.07\mathrm{CF}^{0.5} + \\ &\left(-53.61 - 80.03\mathrm{CF} + 132.78\mathrm{CF}^{0.5}\right)e + \\ &\left(86.61 - 18.91\mathrm{CF} + 163.61\mathrm{CF}^{0.5}\right)e^{0.5}\end{aligned} \tag{7.4}$$

In this equation, k is in square meters and CF is in percent. Table 7.1 shows the range of permeability coefficient values of different intact rocks for comparison.

7.2 Surface permeability (theoretical aspects)

In this section, the concepts and findings related to performing the surface permeability test on an isotropic porous medium are briefly summarized for a better understanding of intact rock permeability. For easy understanding of this experiment, the datum and velocity heads can be neglected in comparison with the pressure load, and Darcy's law can be written as follows (Selvadurai et al. 2020):

$$v(x) = -\frac{K}{\mu}\nabla p(x) \tag{7.5}$$

Table 7.1 Range of the permeability coefficient values for some common rocks

Rock group	*Rock type*	*Permeability coefficient (m/s)*
Igneous	Granit	10^{-14} to 10^{-12}
	Diorite	10^{-14} to 10^{-12}
	Gabbro	10^{-14} to 10^{-12}
	Riolite	10^{-14} to 10^{-12}
	Andesite	10^{-14} to 10^{-12}
	Basalt	10^{-14} to 10^{-12}
Sedimentary	Conglomerate	10^{-16} to 10^{-8}
	Sandstone	10^{-10} to 10^{-8}
	Shail	10^{-13} to 10^{-10}
	Mudstone	10^{-11} to 10^{-9}
	Dolestone	10^{-12} to 10^{-11}
	Limestone	10^{-13} to 10^{-10}
Metamorphic	Gniess	10^{-14} to 10^{-12}
	Schist	10^{-11} to 10^{-8}
	Phylite	10^{-13} to 10^{-11}
	Slate	10^{-14} to 10^{-12}
	Marble	10^{-14} to 10^{-11}
	Quartzite	10^{-14} to 10^{-13}

In this relation, $v(x)$ is the spatially averaged fluid velocity vector with dimension of L/T, K is the isotropic permeability with dimension L^2, μ is dynamic viscosity of water at the test temperature with dimension FT/L^2, $p(x)$ is the fluid pressure with dimension of F/L^2, x is the position vector and ∇ is the gradient operator. Considering the conservation of fluid mass, which applies to an incompressible fluid, the partial differential equation governing the flow is obtained as follows:

$$\nabla^2 p(x) = 0 \tag{7.6}$$

In this relation, ∇^2 is the Laplace operator. When investigating the flow created by the permeameter in the vicinity of the surface of the porous medium, it can be assumed that the dimensions of the flow domain can correspond to a half-space region, especially if the permeameter is placed on the surface of the cuboidal sample and away from its edges. Furthermore, since the sealed region is annular, the stiffness of the subject is reduced to an axisymmetric flow starting from initiated from the annular region. In this case, the relationship 7.6 becomes as follows:

$$\left(\frac{\partial^2}{\partial r^2} + \frac{1}{r}\frac{\partial}{\partial r} + \frac{\partial^2}{\partial z^2}\right) p(r,z) = 0 \tag{7.7}$$

In this relation, r and z are radial and axial coordinates, respectively, in the cylindrical polar coordinate system (r, θ, z). Solutions of the mentioned relation, applicable for axisymmetric problems related to the semi-infinite domain, are as follows:

$$\begin{aligned} & p(r,0) = p_0; 0 \le r \le a \\ & \frac{\partial p}{\partial z} = 0; a \le r \le b \\ & p(r,0) = 0; b \le r \le \infty \end{aligned} \tag{7.8}$$

The problem of the above three-part mixed boundary conditions can be reduced to a set of three integral equations, and the above three-part mixed boundary conditions can be obtained as a series expansion in terms of the parameter $c\ (=a/b)<1$. An acceptable result for the interpretation of surface permeability tests, from performing this test method on cubic samples of Stenstad and Lac du Bonnet granites with a constant flow rate Q_0 (with dimension L^3/T) from the interior region of the impermeable annulus, when the interior region is under constant fluid pressure p_0 and the exterior region is subjected to zero pressure, is obtained. The result can be expressed as follows:

$$Q_0 = ap_0\left(\frac{K}{\mu}\right)F(c) \tag{7.9}$$

In this relation,

$$F(c) = 4\begin{Bmatrix} 1+\left(\dfrac{4}{\pi^2}\right)c+\left(\dfrac{16}{\pi^4}\right)c^2+\left(\dfrac{64}{\pi^6}+\dfrac{8}{9\pi^2}\right)c^3+\left(\dfrac{64}{9\pi^4}+\dfrac{256}{\pi^8}\right)c^4 \\ +\left(\dfrac{92}{225\pi^2}+\dfrac{384}{9\pi^6}+\dfrac{1024}{\pi^{10}}\right)c^5+O\left(c^6\right) \end{Bmatrix}$$

In the limited state when the penetrometer corresponding to a circular entry point with radius a is located on the sealed surface $b \to \infty$ and $c \to 0$, the above relation shows the result of the exact closed-form as follows:

$$Q_0 = 4ap_0\left(\frac{K}{\mu}\right) \tag{7.10}$$

For the aspect ratio of $c=0.25$ related to the geometry of the sealing region, the above relationship will be as follows:

$$Q_0 \simeq 4.458\ ap_0\left(\frac{K}{\mu}\right) \tag{7.11}$$

7.3 Permeability coefficient determination tests

For intact and unfractured rocks, fluid flows through small pores randomly distributed in its matrix. The hydraulic and mechanical behavior of intact rock can be described by continuum mechanics, like soil, which is relatively simple. In this case, the permeability of the porous medium is measured using an empirical relationship, known as Darcy's law, which is related to a fluid with a gradient. The permeability coefficient is measured longitudinally or laterally in intact rock in the laboratory.

7.3.1 *Longitudinal or axial permeability determination tests*

In the longitudinal permeability test, the specimen to be tested is a rock core, in which gas or water flow is applied in the direction of the specimen axis. According to the length and diameter of the specimen, the duration of the fluid passing through the specimen and the difference in the fluid head on both sides of the specimen, its permeability is calculated using the relevant relationships. In soil and rock mechanics, the permeability coefficient is usually measured using falling and constant head tests.

7.3.1.1 *Falling head test*

In this method, a cylindrical specimen of rock is placed in a mold and waxed around it. The mold has a water inlet valve and a water outlet valve. The water inlet valve is connected to a graded pipe, and the water outlet valve is connected to a container where the water level is constant inside it. In this method, after saturating the specimen, the water level in the graded pipe is filled up to the h_1 level and the flow valves are opened. In time t, the water level reaches the lower level h_2. In the falling head test, the water head difference between upstream and downstream of the specimen, always decreases during the test under atmospheric pressure. In this method, the volume of water passing through the specimen is calculated from the following equation (quoted from Fereidooni 2023):

$$q = K.A.i.t \tag{7.12}$$

In this relation, $q = a \times dh$ is the volume of water passing through the specimen, K is the permeability coefficient, A is the cross-sectional area perpendicular to the water flow, $i = h / L$ is the hydraulic gradient of the system and t is the duration of the test. Also, dh is the change of water head during the test, a is the cross-sectional area of the graded pipe and L is the length of the tested specimen. By replacing the values of i and q in the above relationship and solving the equation, the permeability coefficient of the specimen can be calculated from the following relationship (Zarooei and Fereidooni 2023):

$$K = \frac{\alpha.L}{A.t} Ln\left(\frac{h_1}{h_2}\right) \tag{7.13}$$

7.3.1.2 *Constant head test*

The method of conducting a permeability test with the constant head is similar to the falling head method, with the difference that the height of the water entering the specimen is kept constant and the volume of water passing through it (q) is measured at time t. In this test, the difference in the water head of the inlet and outlet of the specimen is always constant during the test under atmospheric pressure. In this method, the permeability coefficient is calculated from the following equation:

$$K = \frac{q.L}{A.h.t} \tag{7.14}$$

The use of falling and constant head permeability tests at atmospheric pressure is useful for rock specimens that have high permeability. Obviously, in the rock specimens with low permeability, water cannot pass through intact rock specimens under atmospheric pressure. In these cases, three other methods are usually used under pressures higher than atmospheric pressure to measure the permeability of the intact rock, which are the steady-state axial gas flow method, the steady-state axial water flow test and the transient pulse axial water flow test.

7.3.1.3 *Steady-state axial gas flow method*

The steady-state axial gas flow method is a type of constant head test that is suitable for measuring very low rock permeability values. The equipment for this test was made by the French company Me´tro-Mesures. The main components of the test are shown schematically in Figure 7.3.

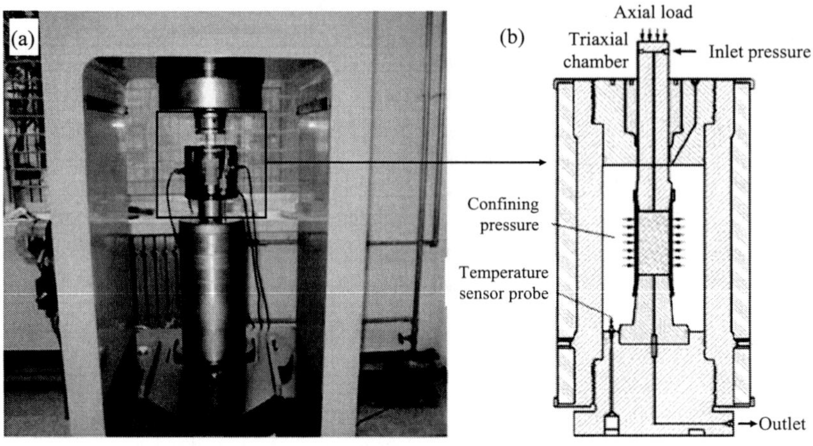

Figure 7.3 (a) Rock permeability testing apparatus for the steady-state axial gas flow method and (b) schematic image of the triaxial chamber of specimen placement in the testing apparatus (Wong et al. 2013).

In this test, nitrogen gas is used as pore fluid. During the permeability measurement test, the inlet gas pressure is kept constant using a gas pressure relief valve, and the gas flow rate at downstream of the specimen is checked. The fluid passing through the specimen is released at its downstream to the atmosphere. The volumetric flow rate of the specimen outlet gas is measured by a bubble gas flow meter with high accuracy. In this method, the permeability of the specimen is determined from the following relationship (Wong et al. 2013):

$$k = \frac{2Q_{v,r}P_r L\mu T_0}{A(p_0 - p_L)(p_0 + p_L)(1+b)T_r} \tag{7.15}$$

In this equation, $Q_{v,r}$ is the volumetric gas outflow rate, measured at output pressure (m^3/s), p_r is the atmospheric pressure (MPa), p_0 is the inlet gas pressure (MPa), p_L is the outlet gas pressure (MPa), T is the specimen temperature in degrees of Kelvin, T_r is the temperature of test room in degrees of Kelvin, μ is the gas viscosity (Pa.s), A is the cross-sectional area of the specimen (m^2), L is the length of the specimen (m), k is the intrinsic permeability of the specimen (m^2) and b is the Klinkenberg coefficient (MPa).

7.3.1.4 *Steady-state axial water flow test*

The method of performing this test is similar to the previous one, except that in this case, water pressure is used instead of gas pressure. During the permeability measurement test in this method, the inlet water pressure is kept constant by a servo-control device, and the outlet water flow rate is monitored at the downstream end of the test specimen by a high-precision flow meter. This test is a type of constant head permeability test. The steps of conducting this test are as follows: first, a cylindrical rock specimen prepared with a transparent rubber sheath is placed in the core holder, and the test temperature is set at about 20°C. Then the required hydrostatic confining pressure (P_c) as well as inlet water pressure (P_f) are applied (Figure 7.4). The inlet water pressure should be applied from the top of the specimen. After that, the output

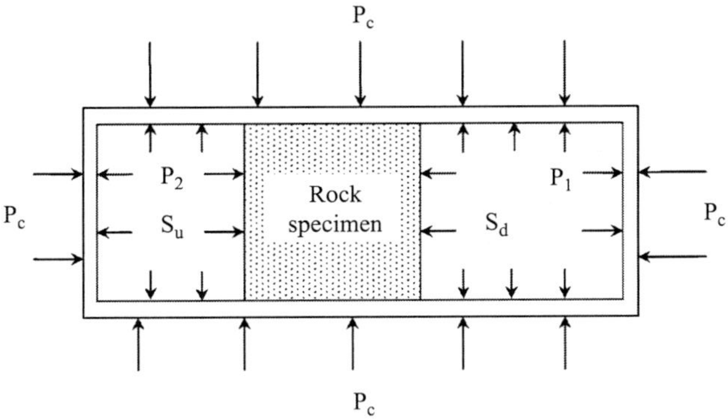

Figure 7.4 A schematic image of rock permeability testing chamber with steady-state axial water flow method (Wong et al. 2013).

water flow rate is measured using a flow meter, and the pressure and flow rate are recorded when the output water flow rate becomes steady. Finally, the hydrostatic confining pressure is changed to a pre-defined value according to a pre-designed sequence (including one-time loading and one-time loading and unloading). Meanwhile, the inlet water pressure is adjusted accordingly and the relevant variables are also recorded. Using the outlet water flow rate (q), the intrinsic permeability of the rock specimen can be calculated based on Darcy's equation as follows (Wong et al. 2013):

$$q = -\frac{k.\partial P}{\mu.\partial X} \tag{7.16}$$

In this equation, q is the volumetric flow of water per unit time and $\partial P/\partial X$ is the pressure gradient. Also, μ is the fluid viscosity with a dimension of FTL^{-2}, and k is the specimen intrinsic permeability with a dimension of L^2.

7.3.1.5 *Transient pulse axial water flow test*

In general, it is not always practical to determine the permeability of very low-permeability rocks using the steady-state fluid flow methods because these rocks require long time to develop a steady flow. Brace et al. (1968) proposed the transient (unsteady) pulse axial water flow method to measure the permeability of a type of granite with very low permeability. This method is based on measuring the depreciation time of hydraulic pressure applied to one end of the specimen instead of determining the velocity or discharge of water in the rock specimen. Therefore, this test is a type of falling head test. An illustration of this method is shown in Figure 7.4. Fluid reservoirs V_1 and V_2 exist at the two ends of the rock specimen with almost the same pressure, so that P_c–P_1 is almost equal to P_c–P_2 (P_c is the pressure of the system enclosure, P_1 is the water pressure in the storage tank 1 and P_2 is the water pressure in the storage tank 2). In this method, to measure the permeability of the rock, first, the pressure P_1 is changed suddenly and by a small amount, and then its return to the initial state over time is considered as the fluid flow between the two reservoirs. Observations during this pressure recovery led to determination of the permeability of the rock specimen. During the test time, the two water storage tanks are connected to the two end surfaces of the rock specimen. By applying an unstable pressure to one end of the specimen and measuring the pressure difference between the two reservoirs, the permeability of the rock specimen can be calculated using the following equations (Wong et al. 2013):

$$\frac{\Delta P(t)}{\Delta P_i} = \exp(-\alpha t) \tag{7.17}$$

$$\alpha = \frac{kA}{\mu L}\left(\frac{1}{S_u} + \frac{1}{S_d}\right) \tag{7.18}$$

In these equations, $\Delta P(t)$ is the measured pressure difference between the upstream and downstream fluid reservoirs in MPa, ΔP_i is the initial differential pressure difference in MPa, t is the measured time in seconds, k is the intrinsic permeability of the specimen in m^2, A is the cross-sectional surface of the specimen in m^2, μ is the dynamic viscosity of water in Pa.s, L is the length of the rock specimen in m and S_u and S_d are the volumes of the upstream and downstream fluid water storage tanks in m^3.

7.3.2 *Transversal or radial permeability determination tests*

The basis of all transversal permeability tests is based on Darcy's law. In these methods, cylindrical rock specimens are used, which have a hole with a certain diameter created along the cylinder axis. The flow of water passing through the specimen is always created in the direction of the radius of the specimen, but it may be in the direction from the inside to the outside of the specimen, or from the outside to the inside (Figure 7.5). The fluid pressure used inside the rock specimen may be one atmosphere or more, depending on the permeability of the rock. Also, the fluid pressure at the inlet and outlet of the specimen may be constant (steady-state) or variable (transient state). In all these methods, the permeability coefficient is obtained from the following relationship, which is based on Darcy's law:

$$K = \frac{Q}{2\pi l \Delta H} . \ln\left(\frac{r_2}{r_1}\right) \tag{7.19}$$

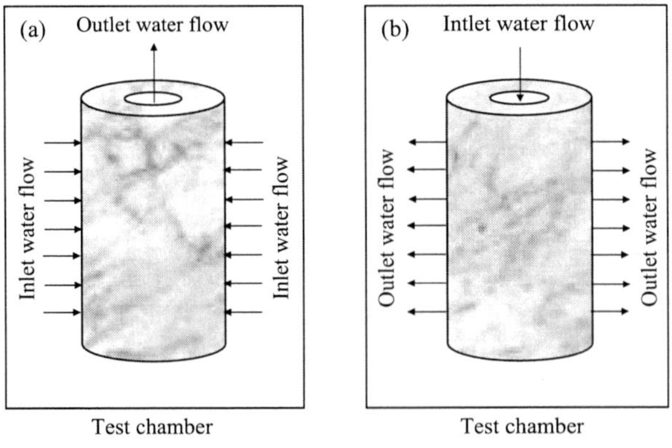

Figure 7.5 A schematic image of the transversal or radial permeability testing chamber of intact rock: (a) flow from outside to inside and (b) flow from inside to outside.

In this equation, Q is the discharge of water passing through the specimen, L is the length of the specimen, r_2 is the radius of the rock core and r_1 is the radius of the hole within it.

7.3.2.1 *Steady-state radial water flow test*

The geometric configuration of the pure radial flow problem involves the solution of the Laplace equation expressed in radial coordinates and subject to boundary conditions. This equation is as follows (Selvadurai and Jenner 2013):

$$p(b) = p_b;\ p(a) = p_a \tag{7.20}$$

In this equation, p_a and p_b are, respectively, the induced pressures in the inner ($r=a$) and outer ($r=b$) parts of a hollow cylindrical rock specimen under the application of steady flow from the inside to the outside Q throughout the length of the cylindrical specimen with a length of H_0 (Figure 7.6). Solving the governing differential equation can be used to develop a relationship, from which permeability can be calculated. This equation will be obtained as follows:

$$k = \frac{Q\mu.\log_e\left(\frac{b}{a}\right)}{2\pi H_0\left(p_a - p_b\right)} \tag{7.21}$$

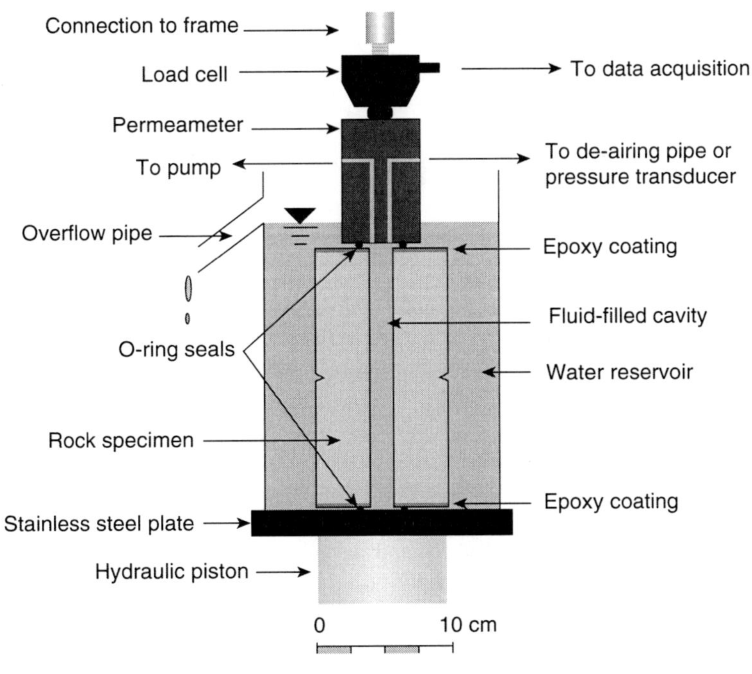

Figure 7.6 A schematic image of rock permeability test chamber with the steady-state radial water flow method (Mattar and Selvadurai 2009).

7.3.2.2 *Unsteady-state radial water flow test*

A suitable analytical approach to solve the partial differential equation of flow is possible only when the flow in a porous medium is completely radial. The analysis of solving the initial boundary value problem governing the dissipation of the pressure pulse from the rock cylindrical hole to the fluid-saturated porous medium of infinite extent can be inspired by the results of the transient heat conduction problem. In this regard, the studies of Selvadurai et al. (2005) provided a complete analysis and showed the range of parameters that ensure the behavior of the flow medium as a fluid-saturated medium with infinite extent. For such a medium, the time-dependent decay of a hydraulic pressure pulse of intensity p_0, applied as the Dirac delta function of time in the central hole of the rock specimen, is obtained from the following equation (Selvadurai and Jenner 2013):

$$\frac{p(t)}{P_0} = \frac{8\alpha}{\pi^2} \int_0^{\infty} \frac{\exp\left(-\frac{\beta u^2}{\alpha}\right)}{uf(u,\alpha)} du \tag{7.22}$$

In this equation,

$$f(u,\alpha) = \left[uJ_0(u) - 2\alpha J_1(u)\right]^2 \left[uY_0(u) - 2\alpha Y_1(u)\right]^2 \tag{7.23}$$

where J_0 and J_1 are the Bessel functions of zero order and first order of the first type, respectively, and Y_0 and Y_1 are the Bessel functions of zero order and first order of the second type, respectively. Also, in the above equation, the dimensionless parameters α and β are obtained using the following relationships:

$$\alpha = \frac{\pi a^2 H_0}{V_w}\left(n^* + \frac{C_{\text{eff}}}{C_w}\right) \tag{7.24}$$

$$\beta = \frac{\pi K H_0 t}{\mu V_w C_w} \tag{7.25}$$

In these equations, V_w is the fluid volume inside the central hole of the cylindrical specimen that increases with increasing pressure and K is the permeability of the porous medium. The other parameters have already been introduced.

References

Brace, W.F., Walsh, J.B., Frangos, W.T., 1968. Permeability of granite under high pressure. *Journal of Geophysical Research*, 73(6), 2225–2236.

Fereidooni, D., 2018. Influence of discontinuities and clay minerals in their filling materials on the instability of rock slopes. *Geomechanics and Geoengineering*, 13(1), 11–21.

Fereidooni, D., 2023. *Engineering Geology*. Damghan University Press, Damghan, 552 pp.

Fereidooni, D., Khajevand, R., (translators), 2019. *Rock Mechanics: An Introduction*. Damghan University Press, Damghan, 280 pp.

Mattar, P., Selvadurai, A.P.S., 2009. Radial flow permeability testing of Indiana limestone. *Proceedings of the 3rd CANUS Rock Mechanics Symposium*, Toronto

Schön, J.H., 1996. *Physical Properties of Rocks: Fundamentals and Principles of Petrophysics*. Pergamon, Oxford.

Selvadurai, A.P.S., Blain-Coallier, A., Selvadurai, P.A., 2020. Estimates for the effective permeability of intact granite obtained from the Eastern and western flanks of the Canadian shield. *Minerals*, 10(8), 667.

Selvadurai, A.P.S., Boulon, M.J., Nguyen, T.S., 2005. The permeability of an intact granite. *Pure and Applied Geophysics*, 162, 373–407.

Selvadurai, A.P.S., Jenner, L., 2013. Radial flow permeability testing of an argillaceous limestone. *Groundwater*, 51(1), 100–107.

Wong, L.N.Y., Li, D., Liu, G., 2013. Experimental studies on permeability of intact and singly jointed meta-sedimentary rocks under confining pressure. *Rock Mechanics and Rock Engineering*, 46, 107–121.

Yang, Y.L., Aplin, A.C., 2010. A permeability–porosity relationship for mudstones. *Marine and Petroleum Geology*, 27(8), 1692–1697.

Zarooei, F., Fereidooni, D., 2023. Assessing the effect of particle size distribution on permeability of silty-sandy soils. *Geotechnical and Geological Engineering*, 6, 1–18.

Zhang, L., 2013. Aspects of rock permeability. *Frontiers of Structural and Civil Engineering*, 7, 102–116.

Chapter 8

Expansion and swelling

8.1 Introduction

Depending on mineralogical composition, rocks can swell in the presence of water, and deformations caused by the increase and decrease in volume cause external pressures. The swelling process can also significantly reduce the shear strength and stiffness of the rocks. In implementation of engineering projects such as tunnels, foundations and roads, the presence of swelling rock at the site can cause great deformations. Arching and applying excessive stress to lining and support system in tunnels, cracking of underground spaces, rise of structure floor, cracking of road and its pavement, landslide and slope instability and destruction of structures located on the swelling rock due to the occurrence of non-uniform settlements are some examples of the swelling rock effects on engineering structures that can lead to severe damage (Fereidooni et al. 2019). Therefore, awareness of the potential of rock swelling is necessary for the proper design and implementation of the structure in the mentioned projects. The term swelling potential refers to the ability of a rock to increase in volume and exert swelling pressure on the surrounding environment. The determining methods of the swelling potential are performed through laboratory tests on intact rock specimens.

8.2 Swelling mechanism

Rock swelling can occur under a single or a combination of several different mechanisms. The mechanism that causes rock swelling depends on several factors, the most important being the mineralogical composition of the rock. The swelling mechanism can be divided into two physical and chemical ones. The physical mechanisms are controlled by electrical forces or pore water pressure, while the chemical mechanisms are always associated with chemical reactions. An overview of the most common swelling mechanisms is provided in Table 8.1.

Pure mechanical swelling caused by negative pore pressure reduction can occur in any argillaceous rock. However, the swelling potential in this mechanism is usually low and less important for tunnel engineering and foundation construction. During microscopic swelling, dipolar water molecules are adsorbed between the clay sheets (Figure 8.1a). The microscopic swelling occurs only when water is absorbed by dry clay minerals such as montmorillonite. This mechanism has a high swelling potential and is observed, for example, in dense bentonite. During osmotic swelling, dipolar water molecules are placed in the dispersed layer of each clay sheet (Figure 8.1b), and if the sheets prevent the movement of water, a repulsive force is created between two adjacent clay sheets (Figure 8.1c).

DOI: 10.1201/9781003672920-8

Table 8.1 An overview of swelling mechanisms (Pimentel 2015)

Swelling mechanisms		*Description*	*Example*
Physical mechanisms	Microscopic swelling	Absorption of the first layer of water between and on the clay sheets	Hydration of very dense dry bentonite
	Osmotic swelling	Adsorption of water molecules in dispersed clay sheets	Hydration of argillitic rocks
	Pure mechanical swelling	Reduction of negative pore water pressure	Hydration of dry or semi-saturated clay
Chemical mechanisms	Converting anhydrite to gypsum	$CaSO_4 + H_2O \rightarrow CaSO_4.H_2O$	Hydration of anhydrite rock
	Oxidation and swelling	Oxidation of pyrite: $FeS + H_2O + 7O \rightarrow FeSO_4.H_2SO_4$ Reaction of anhydrite with carbonate and swelling: $H_2SO_4 + CaCO_3 + 2H_2O \rightarrow CaSO_4.2H_2O.H_2CO_3$	Hydration-drying cycles of pyrite-bearing claystones

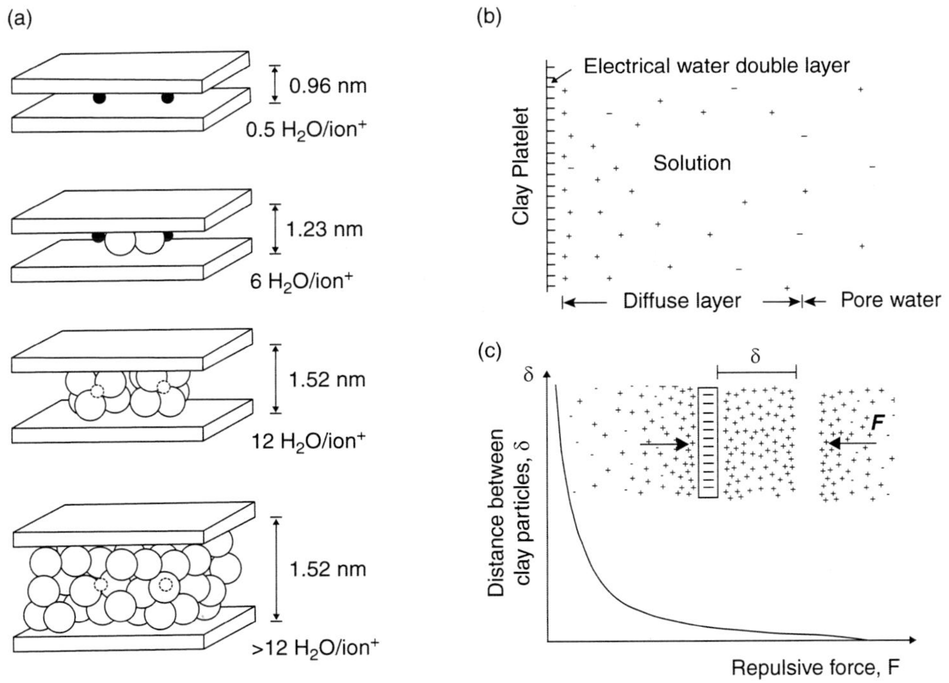

Figure 8.1 (a) Microscopic swelling: expansion of the smectite structure due to water absorption; (b) osmotic swelling: the electric and water-distributed two-layer model for a clay sheet; and (c) theoretical relationship between the repulsive force and distance between two clay sheets (Pimentel 2015).

During the conversion of anhydrite to gypsum, anhydrite dissolves and gypsum precipitates due to the oversaturation of sulfate ions in pore water. In this way, the increase in solid volume reaches 61%. Depending on rock type, the conversion of anhydrite to gypsum often occurs along with osmotic swelling. Swelling caused by pyrite oxidation has been observed beneath some warm deposits where hydration–drying cycles occur due to seasonal fluctuations of groundwater. In tunneling, road construction and foundation engineering, the most common mechanisms of swelling include osmotic swelling and the conversion of anhydrite to gypsum (Diamantis and Fereidooni 2023).

In summary, the most important factors that can affect the swelling behavior of rock are:

- Mineralogical composition, including the amount of swelling and non-swelling clay minerals, anhydrite or pyrite
- Microstructures in the rock (caused by the distribution of minerals)
- Large structures in the rock (as water-permeable joints)
- Chemical composition of water (presence of sulfate or other salts in water)
- External loads and the presence of water (overburden and reinforcing structures)
- Temperature (the effect of temperature is usually negligible. It can only be effective at high temperatures during chemical swelling due to changes in solubility).

8.3 Specimen preparation

To perform free swelling tests, the specimen can be cut with an ordinary rock cutting saw, but for oedometric tests, the specimen preparation process is more difficult, as the specimen must be well fixed in the oedometer ring. For rocks that can be easily broken and crushed, the International Society of Rock Mechanics (ISRM 1999) recommends using a manual press and a specimen-making ring to prepare the specimen (Figure 8.2a). In the recommendations of DGGT (1986), the use of a lathe is suggested to cap the rock specimen in the sampler (ring) of the testing apparatus (Figure 8.2b). VSS (2006) recommends smaller diameter core drilling in the laboratory (overcoring) for cases where the initial core from drilling is not perfectly cylindrical.

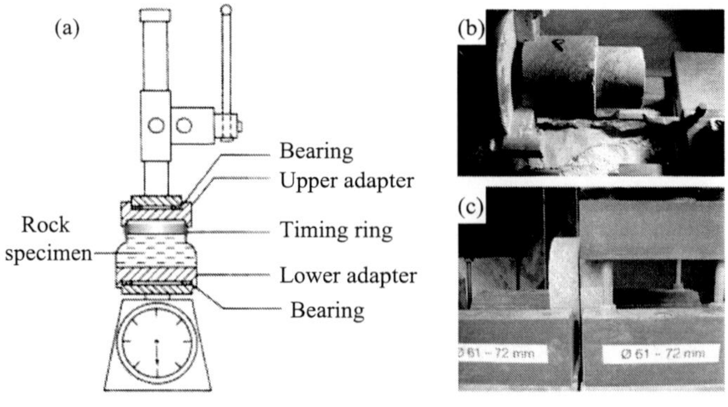

Figure 8.2 Different specimen preparation methods for swelling tests: (a) specimen preparation with a manual press and a manufacturing ring (ISRM 1999), (b) specimen preparation with a lathe and (c) specimen preparation by cutting both ends of the specimen with a diamond band saw (Pimentel 2015).

Obviously, if no overcoring is done, a new metal ring and filter plates must be made for each test, because core dimensions can vary from one drill bit to another. All specimen preparation steps are done with air washing to prevent premature swelling process.

8.4 Swelling determination tests

The purpose of providing different test methods to determine the swelling of rock is to provide information to evaluate the swelling potential of intact rock and ideally in the form of a stress–strain relationship, which is known as the law of swelling and is the only accepted law for pure argillaceous rocks. Various swelling test methods for intact rock samples have been developed since the 1970s. The first proposed method for conducting swelling tests was published by the International Society of Rock Mechanics in 1979 and updated in 1989 and 1999 (ISRM 1979, 1989 and 1999). The latest update only included a few minor changes. In Germany, recommendations were published by the DGGT (now Deutsche Gesellschaft für Erd-und Grundbau, DGEG) in 1986. The Swiss recommendations published in 2006 by the Schweizerischer Verband der Strassen- und Verkehrsfachleute (VSS) are also based on the ISRM (1999) recommendations.

The swelling determination tests recommended in the standards for rock can be classified into two general groups: swelling tests under unconfined conditions (free or unconfined swelling test) and tests under oedometric conditions (semi-confined or confined), namely, radially confined tests. The oedometer tests can be performed to determine maximum swelling pressure, maximum swelling strain or swelling strain under various loads. These tests are classified as one-dimensional tests and can be performed with standard oedometers used in soil mechanics to perform consolidation tests. For all tests, the swelling process is started by immersing rock specimen in water.

8.4.1 *Free swelling strain test*

In general, the main advantage of the free swelling tests is rapid preparation of the test specimen. The maximum swelling strain can be measured by submerging a rock specimen under unconfined conditions. In the initial version of the standard method proposed by the International Society of Rock Mechanics (ISRM 1979), a cube specimen is prepared from a cut rock and deformation of the specimen after submerging is measured in three perpendicular directions (Figure 8.3a). But in a later version (ISRM 1999), it is suggested to test a disc-shaped rock specimen (Figure 8.3b). During the test, the axial deformation of the specimen is continuously measured with a gauge, while the radial deformation is measured with a stainless-steel ring when the test is completed.

After performing the free (unconfined) swelling strain test, the value of the unconfined swelling strain of the rock specimen is obtained from the following equation:

$$\varepsilon_s = \frac{\delta}{L} \times 100 \tag{8.1}$$

In this equation, ε_s is the index of unconfined swelling strain along one of the three spatial axes, δ is the maximum amount of swelling recorded along the desired length in mm and L is the initial distance between the measurement points along the desired length (specimen dimension) in mm.

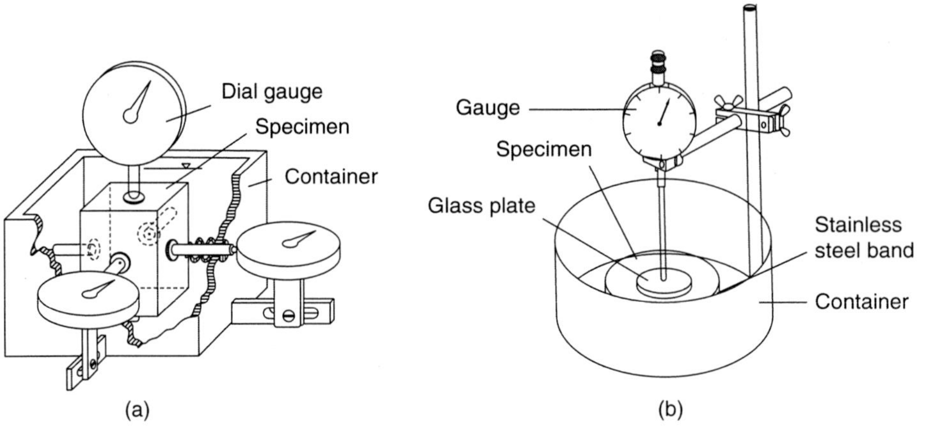

Figure 8.3 Performing methods of the free swelling tests based on recommendations of (a) ISRM (1979) and (b) ISRM (1999).

A potential problem with the free swelling testing is that the specimen under the test can undergo deterioration and slaking, leading to questionable measurements. In addition, measuring the radial deformation with a metal tape (Figure 8.3b) is not very conclusive. Usually, all deformation measurements of an apparatus should have the same precision. In this case, the measuring gauge continuously measures with an accuracy of 0.01 mm or better (the accuracy of standard digital gauges is 0.003 mm), while the metal ring can be read at the end of the test, and this method is used on a semi-deteriorated specimen. It has an accuracy of 0.05 mm. The test method proposed by ISRM in its first recommendations (Figure 8.3a) solved the problem of deformation measurement accuracy using additional measuring gauges placed in orthogonal directions. This method allows the swelling deformation to be measured, that is, perpendicular and parallel to the plane of the bedrock and thus the strain anisotropy to be quantified, provided that the specimen does not undergo deterioration. In the case of anhydrite, which is a soluble rock, as a result of the dissolution of the specimen, sulfate ions are released into the water and precipitated as needle-shaped crystals of gypsum on the free surfaces of the specimen, which cannot be considered as swelling of the rock (Fereidooni 2022).

8.4.2 Oedometric swelling tests

Only the DGGT standard (1986) recommends using the oedometric test to determine the maximum swelling pressure. For this purpose, a relatively low dead load is applied on the specimen (e.g. 5 kPa) in the oedometer and the axial deformation is recorded after the specimen is submerged. Similar to the free swelling tests, no chamber is required to perform this test. This test is known as the semi-confined swelling test. But in another supplementary test, after determining the maximum swelling strain that was described, the oedometer with the specimen is placed in a test chamber and gradually loaded so that the deformations measured due to swelling return to the initial state. The maximum stress required for this purpose must match the maximum swelling stress. This test is known as the confined swelling test.

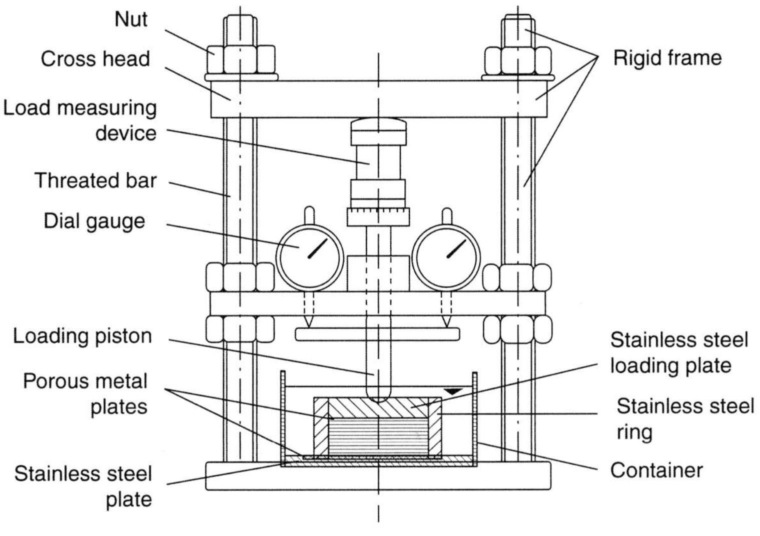

Figure 8.4 A schematic image of the maximum swelling stress measuring apparatus (swelling pressure test in confined conditions) (ISRM 1999).

8.4.2.1 Swelling pressure test in confined conditions

The maximum swelling stress is determined by completely preventing the axial deformation of the specimen and thus achieving the full limit of the specimen volume. A rigid test chamber is required for this purpose. An example of such apparatus is shown in Figure 8.4. The load frame basically consists of two threaded rods on one plate and a series of crossbars. Although the frame must be rigid, a slight deformation of the specimen may occur. If this deformation is significant, the deformation must be compensated, that is, the deformation must be reduced to zero by tightening the nuts on the threaded rods.

After conducting the swelling pressure test in confined conditions, the swelling pressure index of the rock specimen is obtained from the following relationship:

$$P_s = \frac{F}{A} \tag{8.2}$$

In this equation, P_s is the swelling pressure, F is the maximum axial swelling force recorded during the test and A is the cross-sectional area of the rock specimen.

8.4.2.2 Swelling pressure test in semi-confined conditions

Since the radial boundary conditions are not oedometric, the results can overestimate the maximum swelling pressure. If the objective is to determine the maximum swelling pressure, it is strongly recommended that a semi-confined oedometric test, also known as the oedometric test, be performed with a relatively low dead load (e.g. 5 kPa), as suggested in DGGT (1986).

With oedometric tests under different axial loads, swelling strain can be determined as a function of axial stress. For this purpose, a load frame is proposed by ISRM (1999), similar to the oedometer used in soil mechanics to perform consolidation tests (Figure 8.5). In this test,

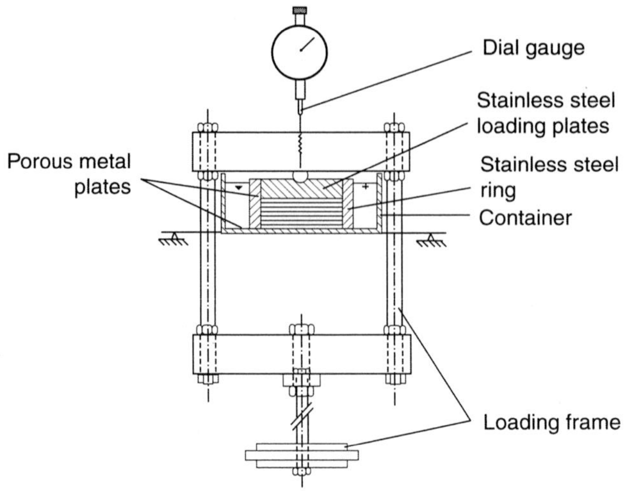

Figure 8.5 A schematic image of the swelling strain measuring apparatus under constant axial load (swelling pressure test in semi-confined conditions) (ISRM 1999).

several load steps are usually applied, that is, different axial loads are applied over time to obtain the stress–strain relationship. Unlike the consolidation tests, the maximum load is applied before submerging the specimen and is kept constant during its swelling, until a new equilibrium is reached and there is not observed further change in axial deformation. After that, the specimen is loaded up to a specified load, and this load is again held constant until a new equilibrium is reached. These steps are repeated for other smaller loads as well. There are two different methods to perform this test, namely, as originally proposed by DGGT (1986) and ISRM (1999). In the first case, the specimen is loaded, unloaded and again loaded until the same load is applied, and then it is submerged, while in the second case, the specimen is loaded only once before submerging.

After performing the swelling strain test in semi-confined conditions, that is, the radial swelling strain test under axial pressure, the value of the semi-confined swelling strain of the specimen is obtained from the following equation:

$$\varepsilon_s = \frac{\delta}{L} \times 100 \tag{8.3}$$

In this equation, ε_s is the semi-confined swelling strain index, δ is the highest recorded swelling value in mm and L is the initial length of the specimen in mm.

After conducting oedometric tests, the water content as well as the density of the specimen is usually determined. The density of the rock grains can be determined in advance with the remaining material from the specimen. For soluble rocks such as anhydrite, it is recommended to dry the specimen in a desiccator over P_2O_5 to determine the water content (ISRM 1999). Then the specimen is heated for 24 hours in an oven with a temperature of 200°C so that gypsum transforms to anhydrite. By weighing the specimen after heating, the amount of water lost is supposed to be determined with sufficient accuracy to calculate the amount of gypsum in the specimen.

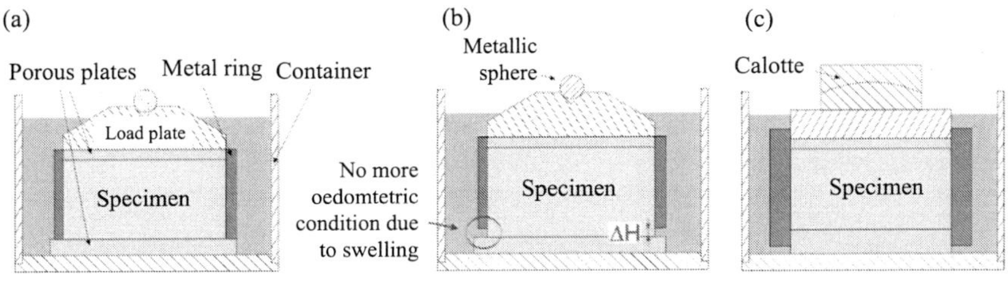

Figure 8.6 (a) Oedometer setup based on ISRM (1999), DGGT (1986) and VSS (2006); (b) possible ring movement due to specimen swelling; and (c) alternative setup.

8.5 Odometer setup

In the study of rock swelling, oedometer setups are taken from soil mechanics based on consolidation tests. In the oedometric test, considering the specimen settlement can be expected, the porous and loading plates located on top of the specimen should be able to move inside the sampler ring (Figure 8.6a). In the case of swelling tests, it is expected that there will be uplift of the sampling ring due to swelling of the rock, in which case the oedometric conditions will no longer be valid. The odometer setup can help prevent this situation (Figure 8.6b). The porous plate at the bottom of the specimen should be placed inside the ring (Figure 8.6c). In addition, a ring with a higher height and a thicker body should be used, because in the case of rocks such as anhydrite, high stresses or large deformations can be expected (Figure 8.6c). Also, the metal sphere can be replaced with a much stiffer metal calotte.

8.6 Odometric test apparatuses

In the recommendations of DGGT (1986) and ISRM (1999) mentioned above, two different types of apparatuses are suggested and required for oedometric tests: one type to determine the maximum swelling stress (Figure 8.4) and the other type to determine the stress–strain relationship of the rock specimen (Figure 8.5). With the first apparatus, deformation-controlled tests are performed, while with the other, only load-controlled tests are possible. In fact, the so-called stiff frame presented in Figure 8.4 is not stiff enough (e.g. even for an inflation stress of 0.5 MPa, an increase in frame stiffness is required). It should be noted that the swelling stresses of anhydrite have been reported to be higher than 8 MPa (Pimentel 2015). Also, increasing the strain rate can damage the specimen. For example, the crushing of gypsum crystals or clay matrix in the rock sample causes maximum false swelling stress. The apparatuses used for consolidation tests in soil mechanics to determine the stress–strain relationship are usually designed for nominal loads of about 10 kN, which may not be efficient in the case of rocks (Khodadad-Zadeh et al. 2022). Apart from this limitation, the maximum load applied during this test should not exceed the maximum swelling stress. Since this value is usually unknown before the test, the applied load is often defined as a function of the overburden pressure in place of the rock and is therefore an arbitrary value. Excessive load can damage or compress the structure of the specimen, thereby questioning the accuracy of the results. This is especially true for weak rocks or for hard soils.

An alternative test method has been presented by Pimentel (2007) at the University of Karlsruhe in Germany, which is a combination of both tests described above. In this method, the specimen is fixed in a rigid chamber and submerged. The confinement of the specimen is maintained until the maximum swelling stress is reached (σ_{max} in Figure 8.7a). Strain compensation

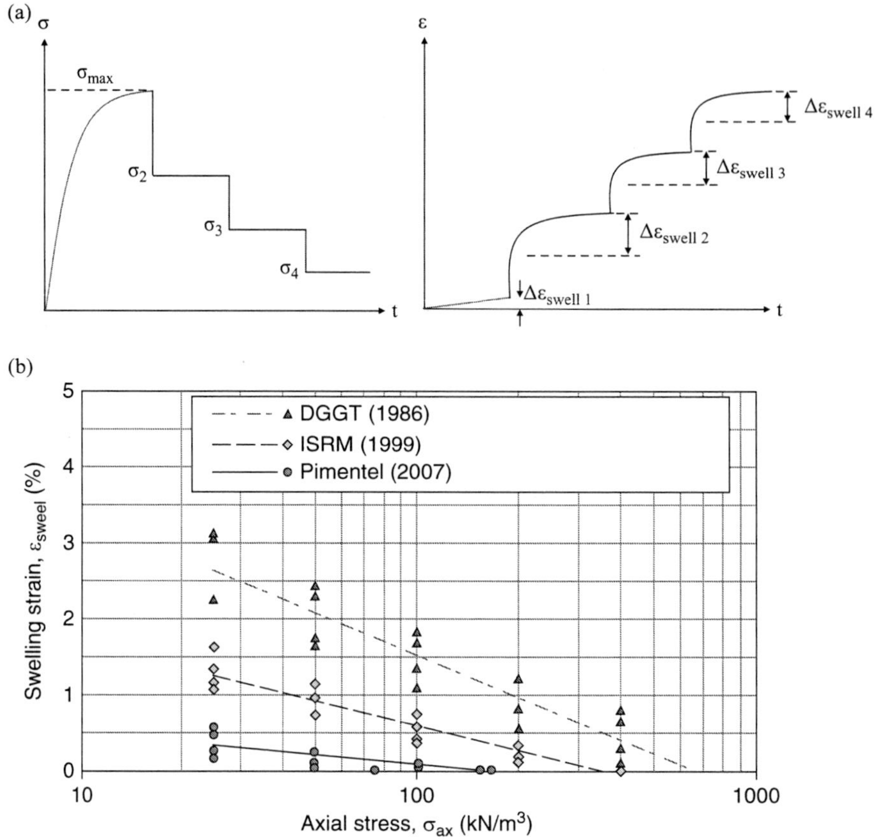

Figure 8.7 (a) Graphs of axial stress and total strain against time in the alternative test method and (b) relationship between stress and swelling strain of Feuerletten samples obtained from three different swelling test methods (Wolffersdorf and Fritsche 2003).

should be avoided. Then, the load is removed from the sample stepwise (e.g. from σ_2 to σ_4 in Figure 8.7a). At each step, the load is kept constant until a new equilibrium is reached. After testing, if necessary, σ_{max} can be corrected by extrapolating the swelling stress–strain curve for zero strain. Figure 8.7b shows the results of the swelling tests in the swelling stress–strain diagram with rock samples from the Feuerletten Formation, drawn using the recommendations of DGGT (1986) and ISRM (1999), and also according to the described alternative method. For this weak rock, it is found that due to preloading, the swelling potential can be significantly overestimated (refer to the value of σ_{max} based on the recommendations of DGGT (1986) and ISRM (1999) and the proposed alternative method that in Figure 8.7b is equal to 610, 330 and 150 kPa). This test method provides the possibility of determining the maximum swelling stress as well as the stress–strain relationship with only one specimen, and even reduces the dispersion of the results without initial loading of the specimen.

The apparatus required for the alternative method described above must be rigid. It should also provide the possibility to perform tests with control of deformation and load and provide more flexibility for test execution. The image of such an apparatus is presented in Figure 8.8.

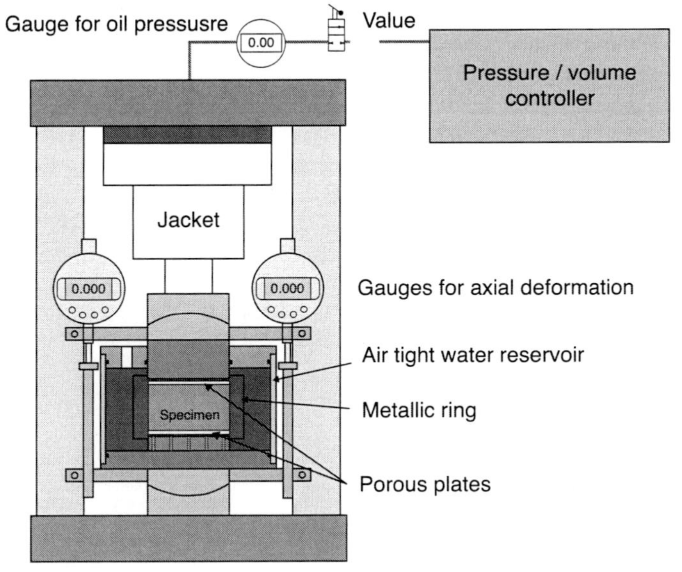

Figure 8.8 Alternative apparatus for oedometric swelling tests (Pimentel 2015).

The reaction frame consists of four columns and two thick steel plates. In the first step, the oil valve is closed in the jacket and the oil pressure is measured with a pressure gauge. For subsequent steps, the oil pressure in the jacket can be controlled with dead loads or alternatively with a pressure/volume controller. To accurately measure the partial deformation of the specimen, two digital measuring gauges should be used as close as possible to the specimen. The water reservoir must be ventilated; otherwise, evaporation will occur. Especially for tests with anhydrite rock, water evaporation changes the ion concentration in the solution and may affect the gypsum deposition process and test results. In the case of anhydrite, the method proposed in ISRM (1999) is not suitable for determining the gypsum content after testing due to lack of accuracy. In this case, the gypsum content can be determined with the dried specimen through thermal analysis as an alternative method. In order to prevent the transformation of gypsum into basanite, the specimen can be dried in a desiccator or in an oven with a temperature lower than 45°C. In addition to the thermal analysis method, X-ray diffraction analysis can be performed before and after the swelling test.

References

DGGT, 1986. Empfehlung Nr. 11 des Arbeitskreises 19 – Versuchstechnik Fels – der Deutschen Gesellschaft für Erd- und Grundbau e.V. *Quellversuche an Gesteinsproben. Bautechnik*, 3, 100–104.

Diamantis, K., Fereidooni, D., 2023. Assessing the Strength and Deformation Properties of Serpentinite Rocks in Dry and Saturated Conditions. *Iranian Journal of Science*, 47(4), 1169–1178.

Fereidooni, D., 2022. Importance of the mineralogical and textural characteristics in the mechanical properties of rocks. *Arabian Journal of Geosciences*, 15(7), 637.

Fereidooni, D., Shirzadipour, K., Abbasi, M., 2019. Assessment of failure reasons of structures in west and southwest of the city of Qom. *Proceedings of the 5th National Symposium of Geology*, Islamic Azad University, Islam-Shahr branch, Islam-Shahr, Tehran, Iran.

ISRM, 1979. Suggested methods for determining water content, porosity, density, absorption and related properties and swelling and slakedurability index properties. *International Journal of Rock Mechanics and Mining Sciences & Geomechanics Abstracts*, 16, 141–156.

ISRM, 1989. Suggested methods for laboratory testing of argillaceous swelling rocks. *International Journal of Rock Mechanics and Mining Sciences & Geomechanics Abstracts*, 26, 415–426.

ISRM, 1999. Suggested methods for laboratory testing of swelling rocks. *International Journal of Rock Mechanics and Mining Sciences*, 36, 291–306.

Khodadad-Zadeh, M., Fereidooni, D., Diamantis, K., 2022. An engineering geological assessment for the Darband dam site, NE of Iran, using eight rock mass classification systems. *Innovative Infrastructure Solutions*, 7(2), 151.

Pimentel, E., 2007. A laboratory testing technique and a model for the swelling behavior of anhydritic rock. *Proceedings of 11th Congress of the International Society for Rock Mechanics (ISRM)*, Lisbon, Vol. 1, pp. 143–146.

Pimentel, E., 2015. Existing methods for swelling tests–a critical review. *Energy Procedia*, 76, 96–105.

VSS, 2006. *SN Schweizer Norm 670 356. Fels Quellversuche*. VSS, Zürich.

Wolffersdorf, P., Fritsche, S., 2003. Laboratory swell tests on overconsolidated clay and diagenetic solidified clay rocks. *Proceedings of an International Symposium*, GTMM, Karlsruhe, pp. 407–412.

Chapter 9

Compressive strength and deformability

9.1 Introduction

Geotechnical engineering projects such as surface and underground excavations and technical structures for detailed analysis and the best design need compressive strength and rock mass deformation parameters. Estimation of rock mass properties, in addition to conducting field tests, can be done through certain experimental relationships and failure criteria using strength and deformability of intact rock. The strength and deformability characteristics check the behavior of rocks against applied force. According to the force nature applied to the rock, these characteristics are different. In other words, the strength and deformability of rocks against forces of different natures are different. Compressive strength examines the rock's response to the application of compressive forces and the maximum amount of force it can withstand until it breaks. The rock behavior during the application of compressive force until the failure moment is associated with deformation in different directions of the rock. This means that various rocks have different strength and deformability properties based on their nature. Therefore, determining the compressive strength and investigating the rock deformation while bearing the compressive force is a subject that will be discussed in detail in this chapter.

9.2 Uniaxial compressive strength

For many years, the uniaxial compressive strength or unconfined compressive strength test has been used as a quantitative laboratory method to determine the strength of intact rock. This parameter is one of the most important mechanical properties of intact rock, which is used in the design, analysis and modeling of rock engineering structures. The results of this test can be directly used for various engineering studies such as tunneling and underground space excavations, slope excavation, foundation construction as well as drilling and blasting in rocks. The test results are also indirectly used to estimate the rock mass behavior, because the uniaxial compressive strength is considered as an important input parameter to describe and classify the rock mass and helps to provide failure criteria for intact rocks.

9.2.1 Measurement of uniaxial compressive strength

Compressive strength is the capacity of a material to resist compressive forces in the direction of the specimen's axis, and it is the most common strength parameter determined for intact rocks. In the laboratory, the uniaxial compressive strength of rocks is obtained in accordance with ISRM (1979) and ASTM-D-2938 (1995a) standards. The recommendation of these standards

DOI: 10.1201/9781003672920-9

Table 9.1 Comparison of ISRM (1979) and ASTM-D-2938 (1995a) standards to determine the uniaxial compressive strength of rock

Parameter	*ISRM standard*	*ASTM-D-2938 standard*
Specimen shape	Cylindrical	Cylindrical
Minimum diameter of specimen (D) (mm)	54	47
Ratio of the height to the diameter of the specimen (*L/D*)	2.5–3	2–2.5
Deviation from the vertical of the end surfaces with respect to the loading axis	0.001 Rad.	0.25°
Maximum roughness of the end surfaces of the specimen (mm)	0.02	0.025
Maximum unevenness of the side surface (mm)	0.3	0.5
Height measurement accuracy (*H*) (mm)	1	–
Diameter measurement accuracy (*D*) (mm)	0.1	–
Diameter of the loading plates (D_p) (mm)	$2 \leq D_p \leq D + 2$	$D \leq D_p \leq 2D$
Maximum thickness of loading plates	15 mm or *D*/3	–
Stiffness of loading plates	58	58
Maximum roughness of loading plates (mm)	0.005	0.013 in new plates, and 0.025 in used pages
Loading rate	0.5–1 MPa/s	–
Specimen failure time (min.)	5–10	5–15
Minimum number of tested specimens	5	10

for preparing a specimen and performing the uniaxial compressive strength test of rocks is summarized in Table 9.1. After preparing the specimen and before performing the test, it is necessary to accurately measure and record the prepared specimen dimensions according to the standards. In this test, a cylindrical rock specimen is subjected to compressive force by an electrical compression jack from its two ends. The loading rate of the specimen is determined according to the standard recommendations so that the specimen finally breaks in a certain time. After the specimen failure, the maximum force endured by it is recorded on the testing apparatus monitor. Figure 9.1 shows an example of uniaxial compressive strength testing machine and its performing mechanism. Also, an image of some tested specimens to understand their failure pattern is presented in this figure. Commonly, the fracture pattern of intact rock specimen in the uniaxial compressive strength test is one or more fracture surfaces diagonal to the loading axis of the specimen. After the test, the value of uniaxial compressive strength is calculated by dividing the maximum compressive force applied on the specimen during the test by its initial cross-sectional area as follows (Fereidooni 2016):

$$\text{UCS} = \sigma_c = \frac{F}{A} = \frac{4F}{\pi D^2} \tag{9.1}$$

In this equation, UCS or σ_c is the uniaxial compressive strength, F is the maximum compressive force applied on the specimen at its failure moment, A and D are the initial cross-sectional area and initial diameter of the specimen, respectively. As mentioned, the uniaxial compressive strength is a very important parameter for intact rock classification. The classification proposed by ISRM (2007) for intact rocks based on the uniaxial compressive strength is presented in Table 9.2.

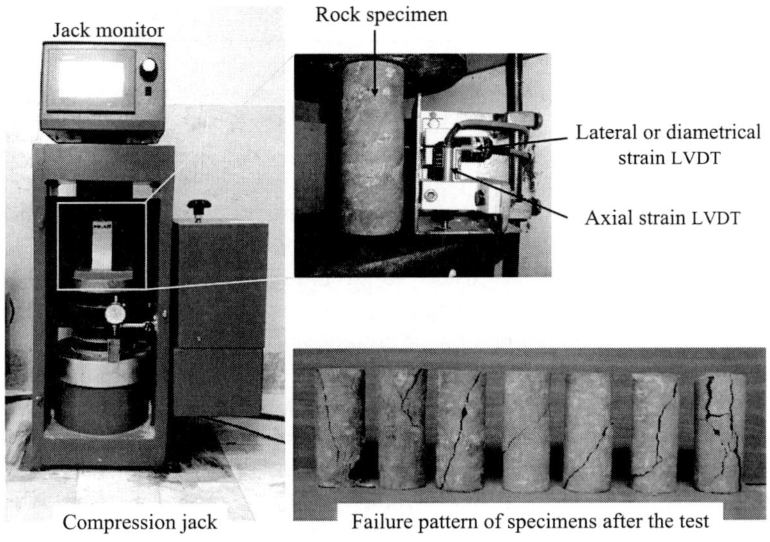

Figure 9.1 Performing mechanism of the uniaxial compressive strength test of rocks and its necessary equipment, along with the failure pattern of some specimens in this test.

Table 9.2 Classification of intact rock based on the uniaxial compressive strength (ISRM 2007)

Uniaxial compressive strength (MPa)	*Strength description*
0.25–1	Extremely weak
1–5	Very weak
5–25	Weak
25–50	Moderate
50–100	Strong
100–250	Very strong
>250	Extremely strong

In conducting the uniaxial compressive strength test of rocks, many factors directly or indirectly affect the test result. These factors can be classified into three groups, including internal or intrinsic, laboratory and environmental factors. The internal factors depend on the intrinsic characteristics of rocks, while the laboratory and environmental factors depend on the test method and its apparatus, conditions of the test environment and the experimenter's skill. Table 9.3 summarizes the most important factors affecting the uniaxial compressive strength of intact rock.

The results of research studies conducted by Fereidooni (2014) and Fereidooni et al. (2016a) indicated that rock anisotropy is one of the most important factors affecting the uniaxial compressive strength of the rock. In anisotropic rocks such as slate, phyllite, schist and hornfels, the minimum uniaxial compressive strength is obtained at anisotropy angles of 15° or 30° with respect to the loading direction of the rock specimen, while the maximum strength is obtained at an anisotropy angle of 90°.

Table 9.3 Factors affecting the uniaxial compressive strength of rock

Factor	*Effect*
Internal or intrinsic factors	• Mineralogical composition: The presence of hard and resistant minerals in rock composition increases its strength (Fereidooni 2022). • Density and porosity: By increasing the density and decreasing the porosity, rock strength increases (Fereidooni and Khajevand 2018). • Size and shape of rock grains: The presence of small and angular grains increases rock strength (Fereidooni and Sousa 2022). • Anisotropy: Anisotropy causes the difference in rock strength in different directions (Fereidooni et al. 2016a).
Laboratory factors	• The effect of end surfaces of specimen and loading plates: The presence of high friction between the end surfaces of the specimen and loading plates increases rock strength. • Specimen size: As the specimen size increases, its strength decreases. • Specimen shape (ratio of height to diameter of specimen, *L*/*D*): The geometric shape of the specimen causes a change in its strength. For this reason, the specimen shape must be according to standard. • Loading rate: Increasing the loading rate increases rock strength. • Rigidity of testing machine: Rigid testing machine increases the accuracy of rock strength determination. • Loading and unloading cycles: By applying loading and unloading cycles, the rock strength is reduced.
Environmental factors	• Moisture content: As the moisture content of rock increases, its strength decreases (Diamantis and Fereidooni 2023). • Pore fluid type: The presence of inactive and dense fluids inside rock porosity increases its strength. • Environment temperature: As the temperature of the test environment increases, rock deformability increases and its elasticity modulus decreases.

9.2.2 *Rock deformability and calculation of static elastic constants*

During the uniaxial compressive strength test, due to the application of axial force, the length of rock specimen gradually decreases while its diameter increases. To measure the strain changes (deformation) of the specimen in the uniaxial compressive strength test, two strain gauges should be installed on the specimen. These gauges measure the specimen deformation in the axial and diametric directions. The length and diameter changes of the specimen can be calculated as follows:

$$\delta = \Delta L = L_0 - L$$

$$\Delta d = d - d_0$$

where δ or ΔL is the change in specimen length; L_0 and d_0 are the initial length and diameter of the specimen, respectively; L and d are the final length and diameter of the specimen; and Δd is the change in specimen diameter. In this case, the relative axial and diametric deformations of the specimen can be calculated through the following relations, respectively:

$$\varepsilon_L = \frac{\Delta L}{L_0} = \frac{\delta}{L_0} \tag{9.2}$$

$$\varepsilon_d = \frac{\Delta d}{d_0} \tag{9.3}$$

After calculating the relative axial and diametric deformations of specimen, its static elastic constants can be calculated. The static elastic constants include elasticity or Young's modulus (E), Poisson's ratio or coefficient (υ), shear modulus or stiffness (G), bulk modulus or incompressibility modulus (K) and Lame's constant (λ). The rock elastic constants can be measured statically or dynamically. The static elastic constants are determined by conducting laboratory tests, while dynamic elastic constants can be determined using the longitudinal and transverse wave velocities in the rock (as explained in the chapter of wave velocity and dynamic elastic constants) (Khajevand and Fereidooni 2018). Definitions and visual concepts of the elastic constants of rock are presented in Table 9.4.

According to the definition, the elasticity modulus is the ratio of stress changes to axial strain changes of specimen during the compressive strength test. The elasticity modulus can be calculated through the following equation:

$$E = \frac{\Delta\sigma}{\Delta\varepsilon} \tag{9.4}$$

Anon's (1979) classification for intact rock based on the elasticity modulus is presented in Table 9.5. Poisson's ratio is the ratio of diametric or lateral strain to axial strain of specimen during the compressive strength test, which is a dimensionless parameter and can be calculated through the following equation:

$$v = \frac{\varepsilon_d}{\varepsilon_a} \tag{9.5}$$

After calculating the elasticity modulus and Poisson's ratio, the shear or rigidity modulus and the bulk or incompressibility modulus are calculated from the following relationships:

$$K = \frac{E}{3(1-2v)} \tag{9.6}$$

$$G = \frac{E}{2(1-v)} \tag{9.7}$$

9.2.3 Stress–strain curve

The result of the uniaxial compressive strength test is a drawing of the stress–strain curve for rocks, and its image is shown in Figure 9.2. This curve consists of the following parts (ISRM 1999):

- Point O: It is the starting point of the test. At this point, both stress and strain are zero.
- Part OA: In this part, the curve has an increasing slope. This means that as the strain increases, the stress increasing rate increases. This issue is due to the closure of the rock specimen porosity.

Table 9.4 Definitions and visual concepts of elastic constants of rock

Elastic constant	*Definition*	*Visual concept*
Elasticity or Young's modulus	Young's modulus is the ratio of axial stress to axial strain of specimen. Axial strain is the change in length of the specimen divided by its initial length. In fact, Young's modulus is the slope of the linear part of the stress–strain curve. Young's modulus and Poisson's ratio are measured in uniaxial compressive or tensile tests.	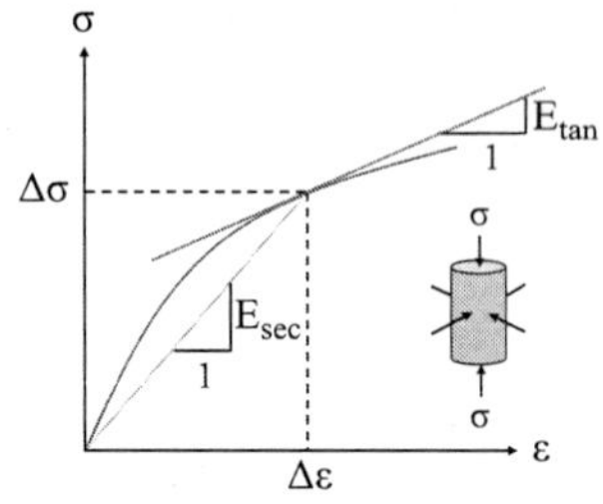
Shear modulus or stiffness	Shear modulus, also known as modulus of rigidity, is a measure of the rigidity of an object, which is obtained from the ratio of shear stress to shear strain.	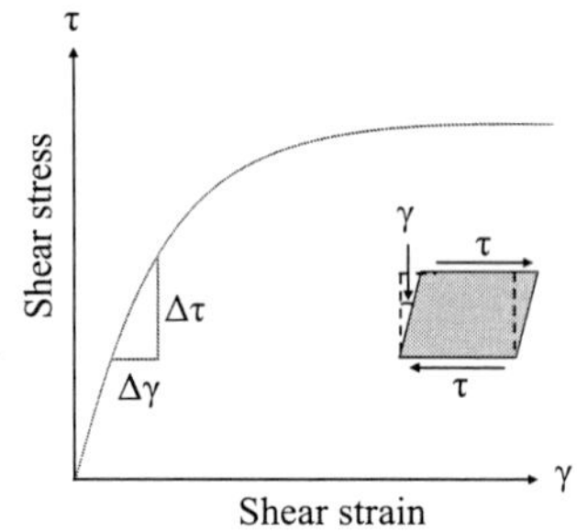
Poisson's ratio or coefficient	Poisson's ratio is the ratio of transverse or lateral strain to the axial strain of specimen, which is represented by a negative sign.	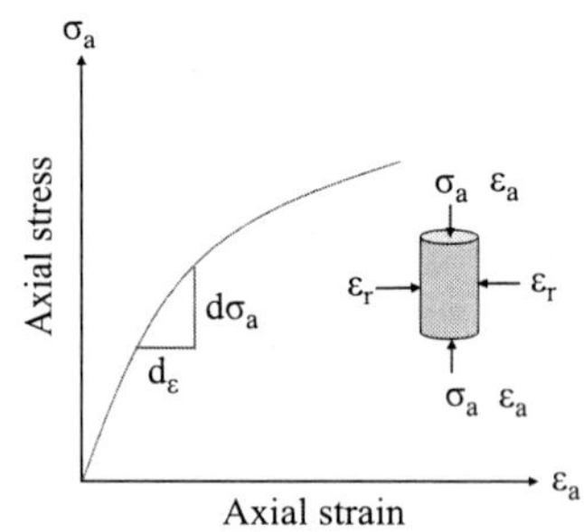
Volumetric, bulk, or incompressibility modulus	Bulk modulus is the ratio of volumetric stress to volumetric strain of rock so that its deformation is within the elastic range. Volumetric or bulk modulus is a numerical constant used to measure and describe the elastic properties of a solid or fluid when pressure is applied to all its surfaces.	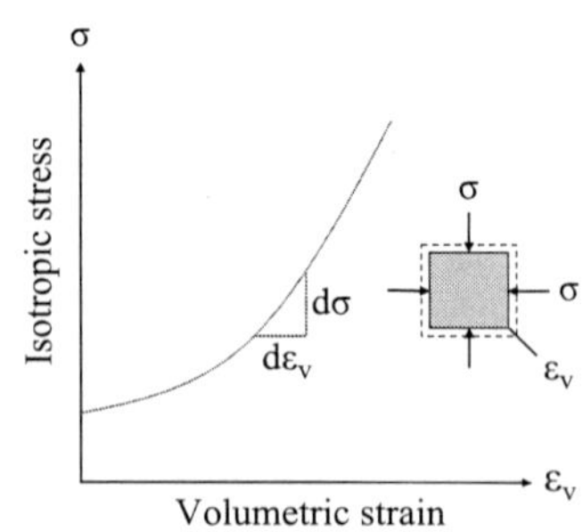
Lame's constant	Lame's constant is the compressive stiffness of the object when the specimen is compressed on one axis and constrained on the other two axes.	–

Table 9.5 Classification of intact rock based on elasticity modulus (Anon 1979)

Class	*Elasticity modulus (GPa)*	*Description*
1	<5	Very low
2	5–15	Low
3	15–30	Moderate
4	30–60	High
5	>60	Very high

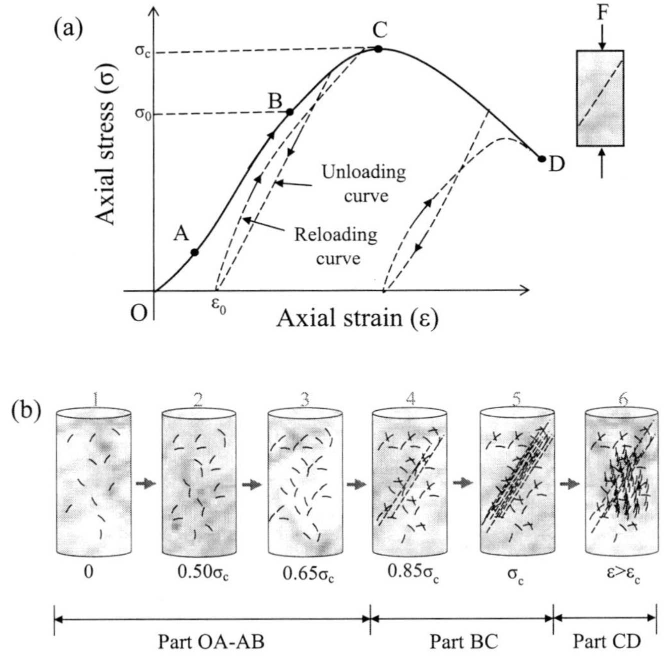

Figure 9.2 (a) General stress–strain curve of rocks and (b) specimen behavior in different stages of loading until the moment of specimen failure, which is associated with an increase in the frequency of microcracks and their orientation along the specimen's failure surface.

- Part AB: In this part, the curve has a constant slope, which means that the rate of strain increase is constant and the behavior of the rock is elastic. The slope of this part of the curve represents the elasticity modulus of rock.
- Point B: This point is called the specimen yield point. At this point, the specimen behavior changes from elastic to plastic.
- Part BC: In this part, the curve has a decreasing slope, and this means that the strain increasing rate increases. The rock behavior in this part is plastic.
- Point C: It is the maximum stress tolerance point of specimen or the peak point. The image of this point on the stress axis is actually the uniaxial compressive strength of the specimen.
- CD part: This part represents the behavior after failure or rupture of the specimen.

After the point of maximum strength, rocks may show one of the following two behaviors (Figure 9.3):

1. Ductile behavior: It refers to the behavior of a rock that, under certain conditions, can maintain a stable shape without losing its resilience under load. This behavior can be seen on the right side of the vertical line passing through the rupture point of the rock in the stress–strain curve.
2. Brittle behavior: It refers to the behavior of a rock, under conditions of which its resilience under load decreases with the increase of deformation. This behavior can be seen on the left side of the vertical line passing through the rupture point of the rock in the stress–strain curve.

One of the important applications of the stress–strain curve is to calculate the elasticity modulus of rock. According to Figure 9.4, the elasticity modulus of rock can be calculated from the stress–strain curve in different states as follows:

- Tangential elasticity modulus: If a tangential line is drawn on the stress–strain curve at any point, the slope of the line is defined as the tangential elasticity modulus at that point. Usually, for comparison, the tangential elasticity modulus at the point of 50% of the final strength of the rock is used.

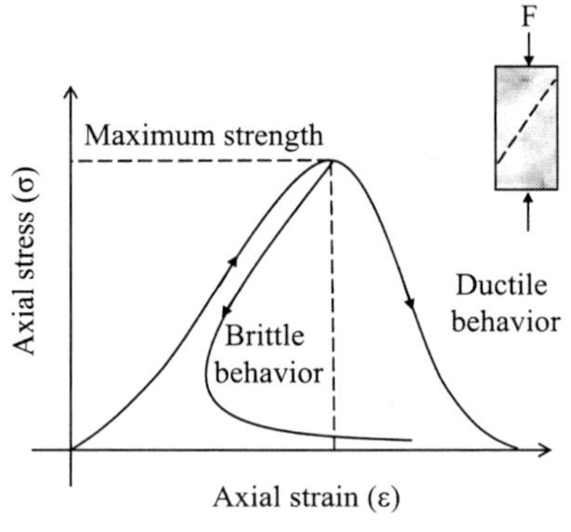

Figure 9.3 Behavior of rocks after the maximum uniaxial compressive strength point.

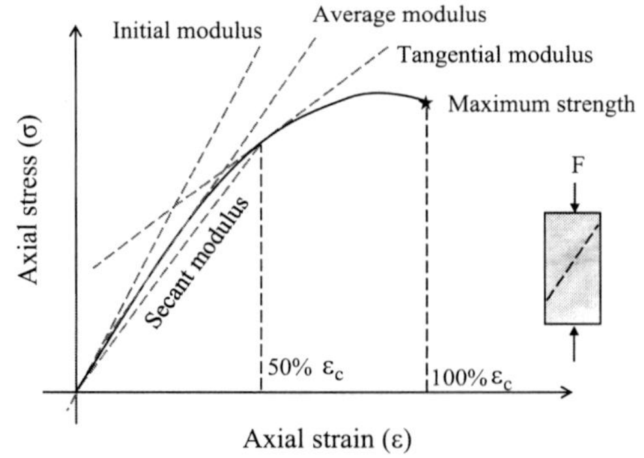

Figure 9.4 Types of elasticity modulus for rocks.

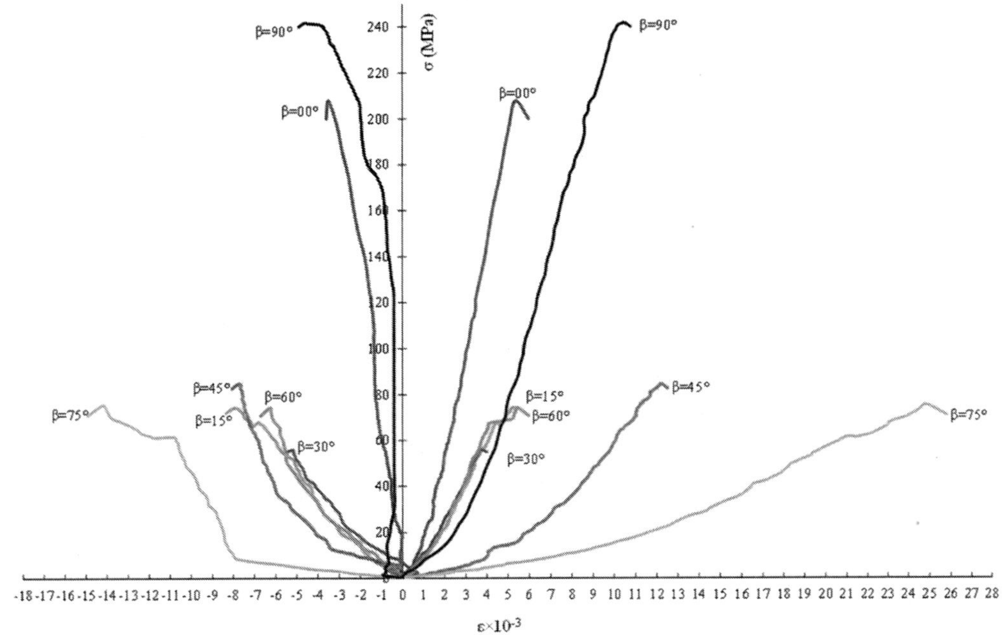

Figure 9.5 Axial and diametric stress–strain curve for an anisotropic hornfels (Fereidooni 2014).

- Secant elasticity modulus: If a line segment is drawn from the origin of the coordinates to any point of the stress–strain curve, the slope of the obtained line is considered as the secant elasticity modulus.
- Average elasticity modulus: It is average slope of the linear part of the stress–strain curve.
- Primary elasticity modulus: If the line segment is tangent to the stress–strain curve from the origin of coordinates, the slope of the obtained line is considered as the primary elasticity modulus.

The stress–strain curve can be drawn for the axial strain and the diametric or lateral strain of rock specimen. For a rock specimen, the axial strain is usually greater than the diametric or lateral strain of the specimen. Axial strain is displayed with a positive sign and lateral strain with a negative sign. Figure 9.5 shows the axial and lateral stress–strain curves for an anisotropic hornfels at different loading angles relative to its anisotropy angle (β). In this figure, the effect of rock anisotropy on uniaxial compressive strength and elasticity modulus is also observed.

9.2.4 *Types of stress–strain curve in intact rock*

The shape of the stress–strain curve is different in various rocks. Based on the shape of the stress–strain curve, all rocks are classified into six different groups according to Figure 9.6. These six groups are:

- Rocks with elastic behavior: The stress–strain curve for this group of rocks is in the form of a straight line with a constant slope that continues to the ultimate strength point. The failure of these rocks occurs explosively. This group includes basalt, diabase, quartzite, dolomite and resistant limestones that have brittle behavior.

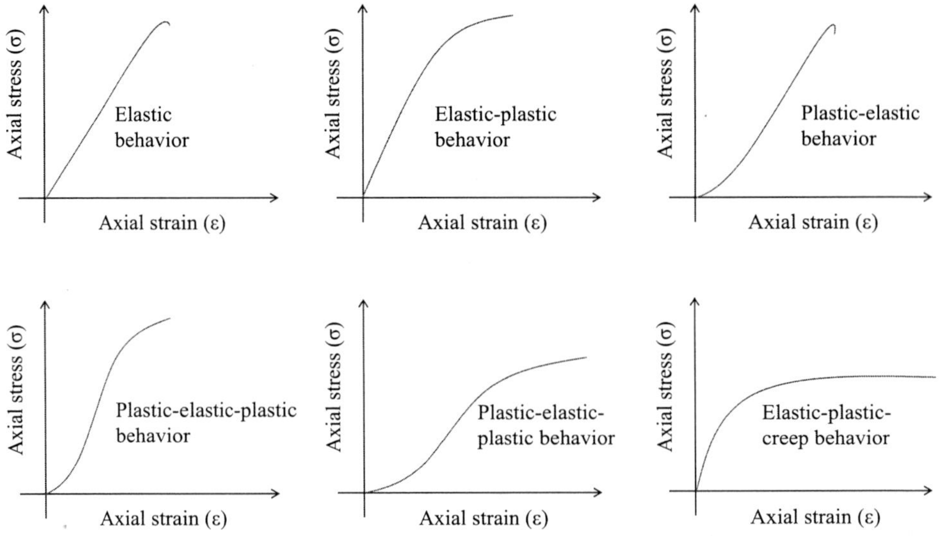

Figure 9.6 Types of the stress–strain curves for intact rocks.

- Rocks with elastic-plastic behavior: The stress–strain curve of this group of rocks is a straight line at first and becomes nonlinear as the loading continues. This behavior is specific to soft limestones and tuffs.
- Rocks with plastic-elastic behavior: The stress–strain curve of this group of rocks is initially curved and becomes linear as the loading continues. This behavior is observed in sandstone, granite and anisotropic rocks such as slate, phyllite and schist, which have been loaded parallel to their anisotropy surfaces.
- Rocks with plastic-elastic-plastic behavior: The stress–strain curve of this group of rocks is stand S-shaped. This means that at the beginning of loading, it is curved, and with the continuation of loading, it takes a linear state and again a curved state. This type of behavior can be seen in gneiss and marble.
- Rocks with plastic-elastic-plastic behavior: The stress–strain curve of this group of rocks is S-shaped. That is, at the beginning of loading, it is curved, and as the loading continues, it becomes linear and curved again. This type of behavior is observed in anisotropic rocks such as slate, phyllite and schist, which are loaded perpendicular to their anisotropy surfaces.
- Rocks with elastic-plastic-creep behavior: The stress–strain curve of this group of rocks is first a straight line, and with continued loading, it becomes nonlinear and finally becomes almost horizontal. This is the special behavior of evaporite rocks such as gypsum rock and salt rock, which exhibit elastic behavior at the beginning of loading, then plastic behavior and finally time-dependent behavior under constant stress.

9.3 Triaxial compressive strength

One of the important tests of rock mechanics to find out the real behavior of the rock against the loads on the site is the triaxial compressive strength test. This test is used to calculate the allowable bearing capacity of rock foundations, dam designs, study of faults and fold mechanisms, and especially underground excavations. By performing the test, the failure envelope of rock will be drawn and strength parameters such as triaxial compressive strength, cohesion and internal friction angle of intact rock are determined. The usual approach for performing triaxial compressive strength test is based on the suggested methods of ISRM (2007) and ASTM-D-2664 (1996) standards. For this purpose, first, a cylindrical rock specimen is prepared according to that was mentioned in the uniaxial compressive strength test. Then the prepared specimen is placed in an insulating cell called the Hoek cell. In the next step, lateral or confining pressure is applied to the lateral surface of the specimen by a hydraulic jack. Then, while keeping the lateral pressure constant, the axial load is applied along the axis of the cylindrical rock specimen, until it breaks (Figure 9.7).

Usually, the confining fluid to apply lateral pressure is hydraulic oil, and the used membrane is oil-resistant rubber. In this test, usually when the applied axial load started to decrease or reached a constant value, or when a certain rock deformation was achieved, the specimen failure occurred. The triaxial compressive strength of a rock specimen under a certain lateral pressure is calculated by dividing the maximum axial force applied to the specimen at the moment of its failure by the initial cross-sectional area of the specimen (Fereidooni 2016):

$$\mathrm{TCS} = \sigma_1 = \frac{F}{A} = \frac{4F}{\pi D^2} \tag{9.8}$$

In this equation, TCS or σ_1 is the triaxial compressive strength of specimen, F is the maximum compressive force on the specimen at the failure moment, A is the initial cross-sectional area of the specimen and D is the initial diameter of the specimen. The amount of triaxial compressive strength is completely dependent on the applied lateral pressure, and a triaxial compressive strength is obtained for each lateral pressure.

Various factors influence the results of the triaxial compressive strength test and the values of cohesion and internal friction angle obtained by it. The most important of these factors are summarized in the following:

- Lateral pressure: The triaxial compressive strength of rock increases with increasing lateral pressure.
- Pore pressure: Increasing pore pressure leads to a reduction in the compressive strength of rock.
- Temperature: Usually, temperature decreases the compressive strength and increases rock deformability.
- Loading rate: Rock strength increases with increasing loading rate.
- Height-to-diameter ratio (H/D) of rock specimen: The final strength of rocks at low lateral pressures is highly dependent on H/D, but at high lateral pressures, it is independent of this ratio.
- Chemical composition of pore fluid: the presence of chemically active fluids reduces rock strength compared to the state in which it is saturated with inactive fluid.
- Anisotropy: In anisotropic rocks such as slate, phyllite, schist and hornfels, the minimum triaxial compressive strength is obtained at anisotropy angles of 15°–45°. While the maximum strength is obtained at an anisotropy angle of zero. In other words, in confined conditions, the minimum triaxial compressive strength value is obtained at different anisotropy angles. So that with the increase of the lateral stress, the anisotropy angle in which the minimum triaxial compressive strength is obtained increases (Fereidooni 2014).

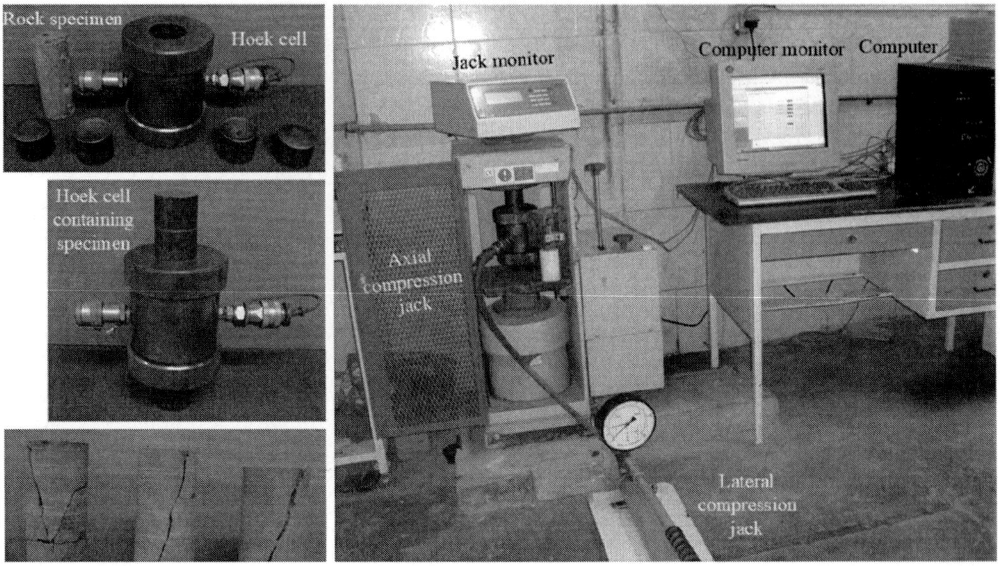

Figure 9.7 Performance mechanism for the triaxial compressive strength test of rocks and its necessary equipment, along with the failure pattern of some specimens in this test (Fereidooni et al. 2016b).

9.4 Point load strength

The point load strength test is used as a standard method to determine the uniaxial compressive strength and classification of intact rock. In addition to the appropriate estimation of the uniaxial compressive strength, this test has the following advantages compared to the uniaxial compressive strength test (Khajevand and Fereidooni 2019):

- In this test, less force is required and a small and portable test device can be used.
- Specimens without a regular geometric shape are used, and there is no need to cut and prepare so many specimens.
- More tests can be performed at the same cost, allowing for adequate sampling when rock conditions are variable.
- Fractured and fragmented rocks can be tested, so there is less chance of biasing the results in favor of intact layers of rock.
- The results of this test show less dispersion than the uniaxial compressive strength test.
- In this test, strength anisotropy measurement is simplified.

Instead, the uniaxial compressive strength test has the following advantages compared to the point load strength test (Jamshidi and Fereidooni 2022):

- The test method is better known and has been evaluated in detail.
- Its results are available for a wide range of rocks, and its results have been correlated with other laboratory and field tests.

Figure 9.8 shows the point load test device during loading in the direction of the axis of a cylindrical rock specimen and the specimen failure pattern after performing the test.

Figure 9.8 Performing mechanism for the point load strength test of rocks and its necessary equipment, along with the failure pattern of some specimens in this test.

The point load strength test is performed in four methods: diametric test, axial test, block test and lump test (specimen without a regular geometric shape). The geometric shapes of the specimen for the point load test are shown in Figure 9.9.

The point load test on lumpy or irregularly shaped specimens was first developed in Russia by Protodyakonov (1960; 1961; 1962). To obtain the strength index, Protodyakonov divided the failure load, which is equivalent to the 2.3 power of specimen volume, by the approximate cross-sectional area of the lumped specimen. The specimen volume was determined using the sand displacement method. The International Society of Rock Mechanics (ISRM 1964) included the Protodyakonov test as a standard method as follows:

Irregular ovoid specimens with a ratio of longest to shortest diameter of about 1.5 to 1 and a volume of about 100 cm^3 are obtained by breaking rocks using any suitable method and rounding suitable specimens with light hammer blows. There should be 15–25 specimens with a mass difference of less than 2%, and these should be broken parallel to their longest axis and perpendicular to the possible heterogeneity surfaces.

In the United States, Reichmuth (1963; 1968) conducted an experimental study of the effects of specimen size and shape on the results of the point load tests on cylindrical and disc-shaped cores. The starting point of his work was to provide a simple formula for determining strength. In subsequent works, this formula was gradually modified in order to fit the curve to the experimental results. His modified formulas, one of which included the concept of compressive strength, were complex and their practical value may be questioned. Following Reichmuth's work, the US Bureau of Mines used the diametric point load test on a core specimen as one of ten index tests in an attempt to predict rock compressive strength. Point load strength by this method provided the highest correlation with the uniaxial compressive strength with a correlation coefficient of 0.947 (Dandrea et al. 1965).

The diametric point load strength test is performed in the most common method on rock cores. Many efforts have been made to generalize the application of this method to specimens

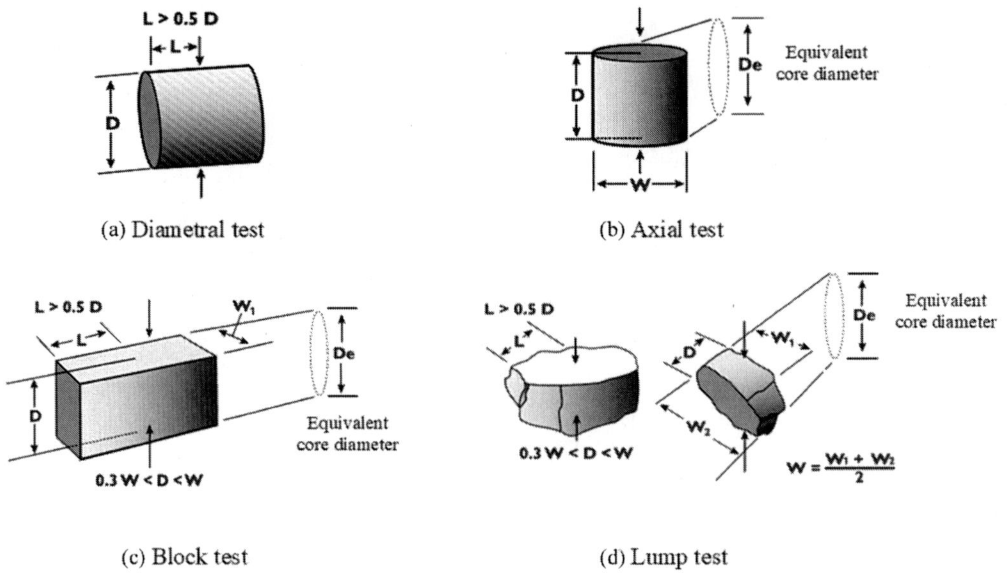

Figure 9.9 Geometric shape of specimens in the point load strength test.

with other shapes. After performing the test, the point load strength index can be calculated by the following equation:

$$I_s = \frac{P}{D_e^2} \tag{9.9}$$

In this equation, P is the maximum force applied to a specimen at the moment of its failure and D_e is the diameter of the equivalent core. One of the applications of the point load strength index is to determine the uniaxial compressive and tensile strengths of rocks. The following relationships can be used to determine the uniaxial compressive strength using the point load index:

$$\sigma_c = \left(14 + 0.175D\right) I_{s(50)} \tag{9.10}$$

In this equation, $I_{s(50)} = F \times I_s$ ($I_{s(50)}$ is the point load strength index of the specimen with a diameter of 50 mm and I_s is the point load index), $F = \left(\frac{D_e}{50}\right)^{0.5}$ (F is the diameter conversion factor) and D is the specimen diameter in mm. The above relationship can be summarized as follows:

$$\sigma_c = 24\ I_{s(50)} \tag{9.11}$$

The following relationship can be used to determine the tensile strength of rocks using the point load test:

$$\sigma_t = 0.94\ I_{s(50)} \tag{9.12}$$

The important point that should be noted in the point load test is that the failure surface must pass through two loading points; otherwise, the test must be repeated. Figure 9.10 shows different valid and invalid failure modes in the point load test.

The point load strength test for rocks is done according to ISRM (1985; 2007) and ASTM-D-5731 (1995b) standards, and the details of these standards are presented in Table 9.6. The point load strength index can be used as a parameter for classification of intact rock and also for classification of rock mass. For classification of intact rock, the point load strength index can be used as presented in Table 9.7. Based on the point load strength index and the uniaxial compressive strength, another classification for intact rocks is presented by Broch and Franklin (1972) (Table 9.8).

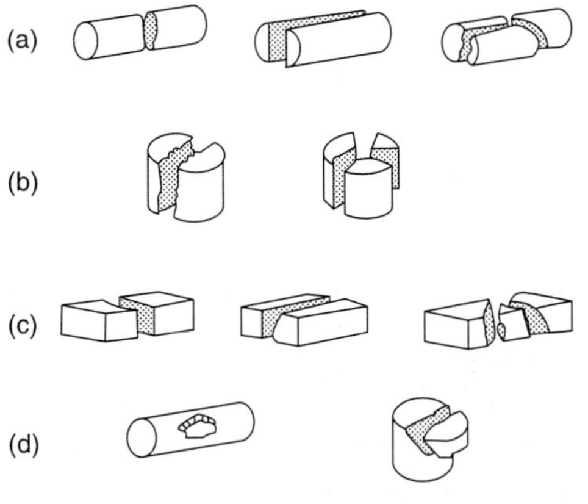

Figure 9.10 Different states of failure in the point load strength test: (a) valid diametric test, (b) valid axial test, (c) valid block test and (d) invalid diametric and axial tests.

Table 9.6 Suggestions of ISRM (1985; 2007) and ASTM-D-5731 (1995b) standards to determine the point load strength index

Test type and specimen specifications	Diametric test: a cylinder with ratio $L/D > 1$ Axial test: a cylinder with a ratio $1/3 < L/D < 1$ Block and lump tests: $1/3 < D/W < 1$ The number of block specimens and cylinders is at least 10 The number of lump specimens is at least 20
Test duration (s)	10–60
Calculations	$I_s = P/D_e^2$ $I_{s(50)} = F.I_s$ $F = (D_e/50)^{1/2}$ For the diametric test: $D_e = D$ For other tests: $D_e^2 = (4WD)/\pi$
Symptoms	D: distance between two jaws (specimen diameter), D_e: equivalent core diameter, L: specimen length, W: specimen width

Table 9.7 Intact rock classification based on the point load strength index

Rock class	*Point load strength index (MPa)*		*Description*
	Deere and Miller classification (1966)	*Bieniawski classification (1989)*	
A or 1	>10	>8	Very high strength
B or 2	5–10	4–8	High strength
C or 3	2.5–5	2–4	Moderate strength
D or 4	1.5–2.5	1–2	Low strength
E or 5	0–1.5	0–1	Very low strength

Table 9.8 Classification of rocks based on the point load strength index (Broch and Franklin 1972)

Point load strength index (MPa)	*Uniaxial compressive strength (MPa)*	*Description*
>10	>160	Extremely high strength
3–10	60–160	Very high strength
1–3	16–60	High strength
0.3–1	5–16	Moderate strength
0.1–0.3	1.6–5	Low strength
0.03–0.1	0.5–1.6	Very low strength
<0.03	<0.5	Extremely high strength

Rock anisotropy is one of the most important factors affecting the point load strength of rocks. In anisotropic rocks such as slate, phyllite, schist and hornfels, the results of point load tests at different anisotropy angles show that as the anisotropy angle increases from 0° to 90°, the values of the point load strength index increase linearly (Fereidooni 2014).

References

Anon, 1979. Classification of rocks and soils for engineering geological mapping, Part 1: Rock and soil materials. Report of the Commission of Engineering Geological Mapping. *Bulletin of the International Association of Engineering Geology*, 19, 364–371.

ASTM, 1995a. *Standard Test Method for Unconfined Compressive Strength of Intact Rock Core Specimens, ASTM Standards on Disc 04.08; Designation: D2938*. ASTM, West Conshohocken, PA.

ASTM, 1995b. *Standard Test Method for Determination of the Point Load Strength Index of Rock. ASTM Standards on Disc 04.08; Designation: D5731-95*. ASTM, West Conshohocken, PA.

ASTM, 1996. *Standard Test Method for Triaxial Compressive Strength of Undrained Rock Core Specimens without Pore Pressure Measurements. ASTM Standards on Disc 04.08, Designation: D2664-95*. ASTM, West Conshohocken, PA.

Bieniawski, Z.T., 1989. *Engineering Rock Mass Classifications*. Wiley, New York, 251 pp.

Broch, E., Franklin, J.A., 1972. The point load strength test. *International Journal of Rock Mechanics and Mining Sciences*, 9, 669–697.

Dandrea, D.V., Fisher, R.L., Fogelson, D.E., 1965. Prediction of Compressive Strength of Rock from Other Rock Properties. U.S. Bureau of Mines, Report of Investigations No. 6702.

Deere, D.U., Miller, R.P., 1966. Engineering classification and index properties for intact rocks. Tech Report No. AFNL-TR, 65-116. Air Force Weapons Lab., New Mexico.

Diamantis, K., Fereidooni, D., 2023. Assessing the strength and deformation properties of serpentinite rocks in dry and saturated conditions. *Iranian Journal of Science*, 47(4), 1169–1178.

Fereidooni, D., Khajevand, R., 2018. Determining the geotechnical characteristics of some sedimentary rocks from Iran with an emphasis on the correlations between physical, index, and mechanical properties. *Geotechnical Testing Journal*, 41(3), 555–573.

Fereidooni, D., Sousa, L., 2022. Predicting the engineering properties of rocks from textural characteristics using some soft computing approaches. *Materials*, 15(22), 7922.

Fereidooni, D., 2014. The study of engineering geological properties of metamorphic rocks in Hamedan province with emphasis on anisotropy. PhD Thesis in Engineering Geology, Bu-Ali Sina University, Hamedan, Iran, p. 224.

Fereidooni, D., 2016. Determination of the geotechnical characteristics of hornfelsic rocks with a particular emphasis on the correlation between physical and mechanical properties. *Rock Mechanics and Rock Engineering*, 49, 2595–2608.

Fereidooni, D., 2022. Importance of the mineralogical and textural characteristics in the mechanical properties of rocks. *Arabian Journal of Geosciences*, 15(7), 637.

Fereidooni, D., Khanlari, G.R., Heidari, M., Sepahi, A.A., 2016a. Assessment of Anisotropy influence on strength and deformability of anisotrpic metamorphic rocks in unconfined condition. *Proceeding of 9th National Conference on Engineering Geology and The Environment*, Kharazmi University, Tehran, Iran.

Fereidooni, D., Khanlari, G.R., Heidari, M., Sepahigero, A.A., Kolahi-Azar, A.P., 2016b. Assessment of inherent anisotropy and confining pressure influences on mechanical behavior of anisotropic foliated rocks under triaxial compression. *Rock Mechanics and Rock Engineering*, 49, 2155–2163.

ISRM, 1964. Richtlinien zur durchfiihrung yon druckversuchen an gesteinen im bergbau, Bericht 5, lander treffen des Internationaten Buros ftir Gebirgsmechanik, Akademie Verlag, Berlin.

ISRM, 1979. Suggested methods for determining the uniaxial compressive strength and deformability of rock materials. *International Journal of Rock Mechanics and Mining Sciences & Geomechanics Abstracts*, 16, 135–140.

ISRM, 1985. Suggested method for determining point load strength. *International Journal of Rock Mechanics and Mining Sciences & Geomechanics Abstracts*, 22, 51–60.

ISRM, 1999. Draft ISRM suggested method for the complete stress-strain curve for intact rock in uniaxial compression. *International Journal of Rock Mechanics and Mining Sciences*, 36(3), 279–289.

ISRM, 2007. Compilation arranged by the ISRM Turkish National Group, Ankara, Turkey. In: Ulusay, R., Hudson, J.A. (eds) *The Blue Book: The Complete ISRM Suggested Methods for Rock Characterization, Testing and Monitoring, 1974–2006*. Kazan Offset Press, Ankara.

Jamshidi, A., Fereidooni, D., 2022. Evaluation of the block punch index test for predicting the strength of sandstones. *Scientific Quarterly Journal of Iranian Association of Engineering Geology*, 15(1), 105–118.

Khajevand, R., Fereidooni, D., 2018. Assessing the empirical correlations between engineering properties and P wave velocity of some sedimentary rock samples from Damghan, northern Iran. *Arabian Journal of Geosciences*, 11, 1–17.

Khajevand, R., Fereidooni, D., 2019. Utilization of the point load and block punch strengths to predict the mechanical properties of several rock samples using regression analysis methods. *Innovative Infrastructure Solutions*, 4, 1–14.

Protodyakonov, M.M., 1960. New methods of determining mechanical properties of rock. *Proceedings of the International Conference on Strata Control*, Paris, Paper C2, pp. 187–195.

Protodyakonov, M.M., 1961. Methods of Studying the Strength of Rocks used in the U.S.S.R. *Proceedings of the International Symposium on Mining Research*, University of Missouri, Missouri.

Protodyakonov, M.M., 1962. Mechanical properties and drillability of rocks. *Proceedings of the 5th Symposium on Rock Mechanics*, University of Minnesota Minneapolis, MN, pp. 103–118.

Reichmuth, D.R., 1963. Correlation of force displacement data with physical properties of rock for percussive drilling systems. In: Fairhurst, C. (ed) *Rock Mechanics* (35–39). Research Laboratory (URL), Pinawa, Manitoba.

Reichmuth, D.R., 1968. Point-load testing of brittle materials to determine tensile strength and relative brittleness. *Proceedings of the Ninth Symposium on Rock Mechanics*, University of Colorado, Boulder, CO, pp. 134–159.

Chapter 10

Tensile strength

10.1 Introduction

Tensile strength is the maximum tensile stress that a material can withstand until it breaks. Knowing the behavior of rocks in tension and their tensile strength is very important for analyzing the stability of roof and floor of underground spaces, investigating the stability of slopes and excavating and blasting in rock (Shariati and Fereidooni 2021; Margan and Fereidooni 2024). Knowing about the tensile strength of rocks is one of the prerequisites for success in designing surface and subsurface engineering structures such as underground and open pit mines, tunnels, slopes and other rock engineering projects. Assessment of rock and its discontinuities in the engineering project construction and knowledge of subsurface layers helps us in understanding tensile failure mechanisms. Rocks and generally all brittle materials are naturally weaker in tension than in compression (Khanlari et al. 2014). Thus, the reason for the occurrence of many failures and instabilities in mines, tunnels, caverns and rock slopes is the creation of tensile stresses (Fereidooni 2004). This phenomenon clearly shows the importance of understanding the issue of tensile strength of rock and mechanism of tensile failure. So far, various methods have been invented to measure the tensile strength of rocks, which can be generally divided into two types: direct tensile tests and indirect tensile tests.

10.2 Direct tensile strength

The best method for measuring the tensile strength of rocks is to pull them directly, which is known as direct tensile strength or uniaxial tensile strength test. The accuracy of this test, which was originally designed for metals, depends on the following conditions:

- The test specimen should be subjected to pure tensile stress only.
- A uniform tensile stress should be applied to the entire cross section of specimen.

In rock mechanics, accurate direct tensile testing is associated with many problems and high costs, which limit its use as a common laboratory method. The main problems of this method include the following (Fereidooni 2016):

- The optimal specimen preparation method will not be simple.
- Connection of specimen to the apparatus jaws is very difficult.
- It is difficult to apply a purely axial tensile load.
- It is possible to create bending and torsional stresses.

DOI: 10.1201/9781003672920-10

In order to eliminate the above problems, many efforts have been made by different researchers and various methods have been presented. In this regard, different geometric shapes of specimens are tested, which include briquette, dumbbell, cylindrical and dog-bone specimens. A schematic image of the specimens is shown in Figure 10.1. Also, Figure 10.2 shows different types of grips used in these methods.

Obert et al. (1946) used cylindrical specimens and grips, as shown in Figure 10.2a. After conducting numerous tests on marble, limestone, granite and sandstone, they reported that this method is not suitable for specimens with a tensile strength higher than 8.27 MPa. Fairhurst (1961) suggested the use of epoxy glue with an approximate tensile strength of 20.69–27.58 MPa for gluing cylindrical specimens to the direct tensile apparatus (Figure 10.2b). This glue establishes a very good connection between specimens and metal rods of the same diameter. This method can be used on prismatic and cubic specimens without any problem. Therefore, it is necessary that the shape and cross-sectional dimensions of the caps are the same as the specimen. This method provides satisfactory results and seems more suitable than other suggested methods. Wuerker (1955) used briquette specimens for the direct tensile test (Figure 10.2c). At first, due to the concentration of high stress in the curvature of the specimen and near the grips, the desired results were not obtained. By making some corrections, he got completely satisfactory results. The biggest problem with this method is specimen

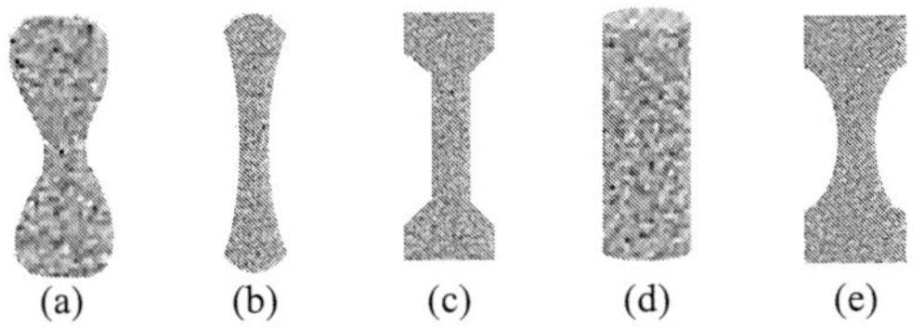

Figure 10.1 Different types of specimens used to determine the direct tensile strength: (a) briquette specimen (Wuerker 1955), (b) dumbbell specimen (Brace 1964), (c) dumbbell specimen (Hook 1964), (d) cylindrical specimen (ISRM 1978) and (e) dog-bone specimen (Townhill-Rewston 1984).

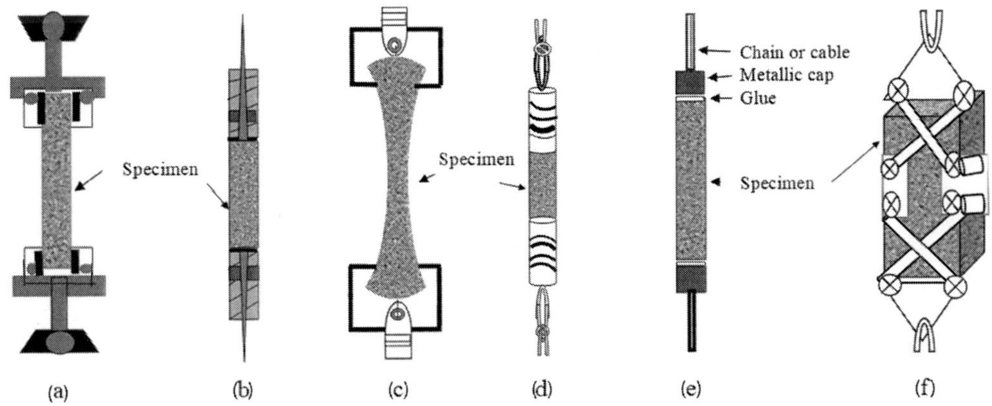

Figure 10.2 Different types of grips used in the direct tensile test: (a) Obert et al. (1946), (b) Fairhurst (1961), (c) Wuerker (1955), (d and e) ISRM (1978) and (f) Townhill-Rewston (1984).

preparation, which is very time-consuming and expensive. ISRM (1978) suggests the use of cylindrical specimens and adhesive in the creation of tensile stresses to perform the uniaxial tensile strength test (Figure 10.2d and e). Townhill-Rewston (1984) also suggested the use of a rectangular cube specimen and mechanical grips to perform the uniaxial tensile strength test (Figure 10.2f).

Currently, the methods provided by ISRM (1978) and ASTM (2020) standards are used to perform the uniaxial tensile strength tests in the laboratory. In this method, the cylindrical rock specimen is affected by the tensile force from its two ends until it breaks. Then, the tensile strength of the specimen is calculated by dividing the maximum load by the initial cross-sectional area of the specimen according to the following equation (Fereidooni 2016):

$$\text{UTS} = \sigma_t = -\frac{F}{A} = -\frac{4F}{\pi D^2} \tag{10.1}$$

In this equation, UTS or σ_t is the uniaxial tensile strength, F is the force applied to the specimen at the moment of failure and A is the cross-sectional area of the specimen. The minus sign is used only to distinguish this test from the uniaxial compressive test, or in other words, to separate the tensile stress from the compressive one. In Table 10.1, the parameters and requirements for conducting the direct tensile strength tests in ISRM (1978) and ASTM (2020) standard methods are compared.

During their studies, Brace (1964) and Hoek (1964) used specimens similar to those used in direct tensile testing of metals (Figure 10.3b and d). These specimens had a cylindrical central part with a length of 4.32 cm and a diameter of 16.2 cm. The radius of curvature was 1.5 cm, and the diameter of the end parts was 1.4 cm. The specimens were tested with conventional edge grips, and satisfactory results were obtained.

Also, Gorski (1993) and Klanphumeesri (2010) performed a direct tensile test on dog-bone specimens with slight changes in the jaws and obtained favorable results (Figure 10.3a). The use of dumbbell-shaped specimens, which was also suggested by other researchers, is accompanied by problems. It will be very difficult to prepare such specimens from brittle materials such as rock because most of rocks are broken and destroyed during coring and preparation stages.

Table 10.1 Comparison of the direct tensile strength test parameters in methods suggested by ISRM (1978) and ASTM (2020)

Parameter	*ISRM (1978)*	*ASTM (2020)*
Minimum cap thickness (mm)	15	30
Minimum specimen diameter (D) (mm)	54	47
Maximum cap diameter (mm)	D+2	1.1 D
Diameter measurement accuracy (mm)	0.1	0.25
Height measurement accuracy (mm)	1	0/25
Ratio of height to diameter of specimen	2/5–3	–
Maximum deviation from vertical end surfaces	0.001 Rad.	0.25 deg.
Maximum roughness of side surfaces (mm)	0.1	0.5
Loading rate (MPa/s)	0.5–1	–
Test duration (min.)	5	5–15
Input force reading accuracy (%)	1	–
Number of tested specimens	5	10
Accuracy of reporting the value of tensile strength	Three decimal places	35 kPa

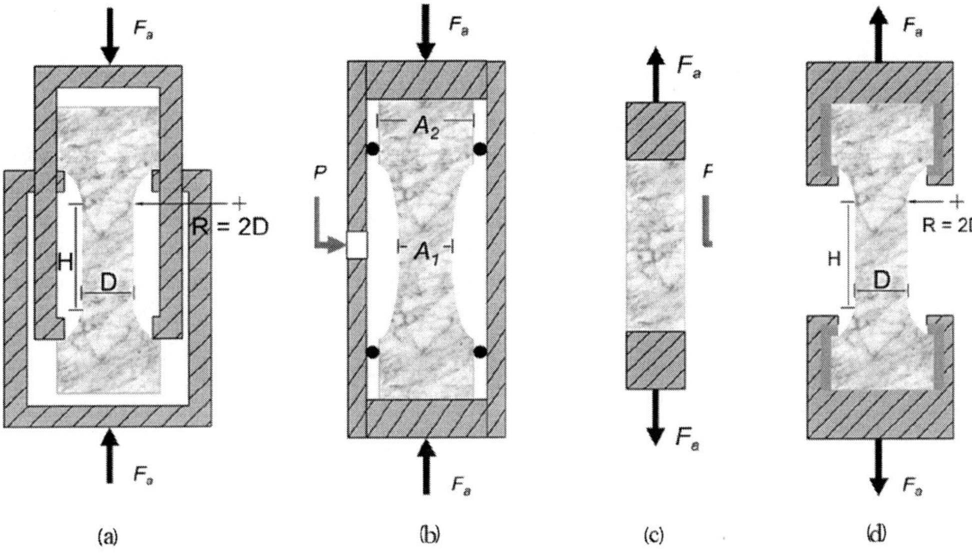

Figure 10.3 Different methods for conducting the direct tensile strength test (adapted from Perras and Diederichs 2014): (a) compressive to tensile load converter, for the direct tensile test (Gorski 1993 and Klanphumeesri 2010), (b) biaxial tensile (Brace 1964), (c) end-glued jaws for cylindrical specimens (ISRM 1978) and (d) griped jaws for dog-bone specimens (described by Hoek 1964). *F* is the applied force, *R* is the specimen radius, *D* is the specimen diameter, *A* is the cross-sectional area of specimen and *P* is the water pressure inside the cylinder.

On the other hand, even in case of success in preparing such specimens, other problems will occur during testing. For example, failure may start from the contact point of the specimen with the apparatus, or if the applied load is not purely tensile, it may cause compressive, bending and torsional stresses.

As mentioned, Townhill-Rewston (1984), in an attempt to solve the problem of specimen preparation, proposed dog-bone shaped specimens. In devising this shape, clay specimens that were traditionally used to measure the direct tensile strength of Portland cement have been adapted. In general, dog-bone specimens have the following advantages compared to dumbbell- and cylindrical-shaped specimens:

- Their preparation is very cheap and simple.
- The stress concentration or end effects of the specimen are not as severe as those of cylindrical specimens.
- Preparation of dog-bone specimens allows for the simultaneous preparation of cylindrical specimens as side products that can be used in other tests.

Despite the many efforts that have been made all over the world to accurately perform the uniaxial or direct tensile strength test, it seems that this test is basically technically difficult and expensive. For this reason, researchers found a great tendency to use indirect methods to determine tensile strength of rocks (Fereidooni and Khajevand 2018).

10.3 Indirect tensile strength

Early attempts to directly measure the tensile strength of rocks and concrete suffered from experimental difficulties due to the stress concentration created by the steel jaws used to grip the specimens. Early failure was common even when special dumbbell-shaped specimens were used. For this reason, various indirect tensile tests were investigated as an alternative to the direct methods for measuring the tensile strength of intact rock. The most important such tests are shown in Figure 10.4. The most well-known of these tests is the Brazilian two-half split test (Figure 10.4a), which was often used in the past to control the quality of concrete and ceramics, and later was also used to test intact rock. Indirect tensile tests can be divided into two general groups, including point load and linear load tests. In these tests, the concentrated vertical applied load causes horizontal tensile stress and the specimen failure occurs by splitting along one or more planes parallel to the load direction. In this regard, elastic stress analysis has been published for various experiments and numerical methods have been used to consider deviations from ideal elastic assumptions. The following general results can be obtained from these analyses:

- The stresses perpendicular to the fracture surface of specimen are tensile type in most points of the failure surface, which are associated with a compressive component in the direction of loading.
- The maximum tensile stress occurring at the center of specimen is related to applied load (P), distance between the jaws (D) and linear loading length (t) through the following equations (Broch and Franklin 1972).

For the linear load test:

$$\sigma_t = \frac{kP}{Dt} \tag{10.2}$$

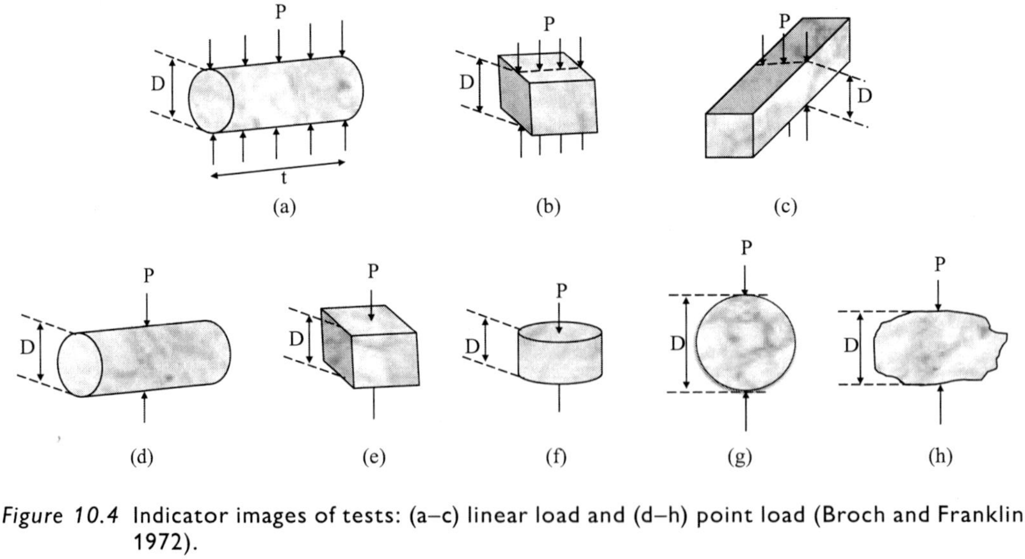

Figure 10.4 Indicator images of tests: (a–c) linear load and (d–h) point load (Broch and Franklin 1972).

For the point load test:

$$\sigma_t = \frac{kP}{D^2} \tag{10.3}$$

The constant k has a value of about 0.5–1 depending on the specimen geometry.

In the indirect tensile tests on rock specimens, the role of compressive force applied to the specimen, which may be greater than the induced tensile stress, cannot be ignored in the results and failure process of the specimen. This problem led to the testing of various alternative forms of indirect tensile testing using disc-shaped specimens with a central hole. In this test, theoretically, there is a pure tension mode above and below the central hole to overcome the problem of compressive stress. But other problems are related to high stress level and stress sensitivity to inelastic behavior. For this reason, the results obtained from different types of indirect tensile tests are actually different from each other and from the direct tensile strength values.

The use of new epoxy glues makes it possible to perform direct tensile testing of prismatic specimens without the previous stress concentration problems, so that indirect tensile tests are now obsolete for measuring tensile strength. However, these tests have an important application in determining the strength index. In this method, many of the mentioned criticisms are no longer included.

The main requirements for performing the tensile strength test are actually the simplicity and repeatability of the test. In this regard, the point load tests are usually less sensitive to irregularities compared to the line load tests. In these tests, the area under critical pressure is located inside the specimen, where surface irregularities have the least effect. In the point load tests, the tensile strength can be expressed as the ratio of the applied force to the square of the distance between the loading jaws. If certain restrictions are imposed on the shape and size of the specimens, the actual geometry of the specimen has little effect on the tensile strength results. This means that the types of point load tests may be chosen to suit the shapes of commonly available specimens, typically either core or lump specimens of irregular shape (Figure 10.4c and h), for which specimens are not cut.

Until now, indirect tensile strength measurement has been done in various methods. These methods are used to indirectly determine the tensile strength of intact rock specimens. Types of indirect tension measurement methods are divided as follows:

- Three-point bending test
- Four-point bending test
- Hydraulic extension test
- Ring tensile test
- Cylinder diametric pressure test
- Sphere diametric pressure test
- Diametric pressure test of square plates
- Centrifuge tension test
- Brazilian strength test

10.3.1 Three-point bending test

One of the common methods for determining the indirect tensile strength of rocks is the three-point bending test. The indirect tensile strength obtained from this method is also called bending strength, failure modulus and maximum tensile strength. Often, the value of this strength is greater than the direct tensile strength. In this test, the specimen is in contact with the test apparatus through three points (Figure 10.5). The specimen is placed on two metal balls that are

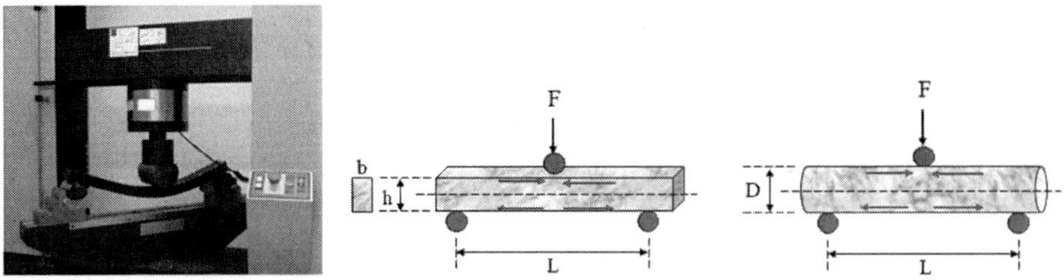

Figure 10.5 (a) Three-point bending test apparatus, (b) performing method of the test on a rock beam with a rectangular section and (c) performing method of the test on a rock beam with a circular section.

placed at a distance L from each other, and the load is applied through another metal ball that is located in the center and above the specimen. This test can be performed on cylindrical, rectangular cube and disc-shaped specimens. In the case of rectangular cube and disk specimens, the supports are in the form of self-adjusting spheres with a completely treated surface. These balls are designed so that they can rotate freely and carry out the load action without creating the slightest deviation in the loading path. In the case of cylindrical specimens, the shape of the supports is not so sensitive and a rod can also be used. The circular and uniform surface of these specimens helps to adjust the supports under the load and prevents stress concentration. Indirect tensile strength in this test is calculated from the main equation of bending strength given in statics, as follows:

$$\sigma_b = \frac{My}{I} \tag{10.4}$$

In this equation, σ_b is the bending strength, M is the bending moment, y is the distance from the neutral axis of the rock beam section and I is the inertia moment or the second order of the rock beam section under test. Indirect tensile strengths in the three-point bending method for rock beams with rectangular and circular sections are calculated from the following relationships.

For a rectangular section:

$$\sigma_t = -\frac{3LF}{bh^2} \tag{10.5}$$

For a circular section:

$$\sigma_t = -\frac{16LF}{\pi D^3} \tag{10.6}$$

In these relations, F is the force at the failure moment; L is the distance between the two supports; b and h are width and height of the specimen in the rectangular section, respectively; and D is the diameter of specimen in the circular section.

10.3.2 *Four-point bending test*

In the three-point bending test, the stress concentration under the loading point causes tensile cracks to appear on the outer surface and just opposite this point, and finally, the specimen is

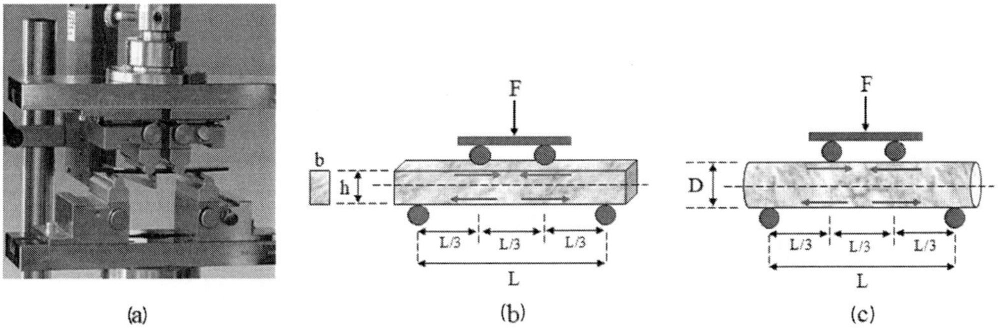

Figure 10.6 (a) Four-point bending test apparatus, (b) performing method of the test on a rock beam with a rectangular section and (c) performing method of the test on a rock beam with a circular section.

nearing failure. Criticisms raised in this field made the four-point bending test to be proposed. In the four-point bending test and in a part of the specimen that is between two loading points, a region with a constant and uniform bending moment appears, and therefore, the specimen is only under pure bending stress in this part (Figure 10.6). This test can also be performed on cylindrical and rectangular cube specimens.

Since in the four-point bending test, a pure bending moment is developed in the distance between the supports; therefore, no special stress concentration is created and failure occurs at the weakest point of this range. For this reason, more accurate results will be obtained than with the three-point method. Indirect tensile strengths in the four-point method for rock beams with rectangular and circular sections are calculated from the following relationships.

For a rectangular section:

$$\sigma_t = -\frac{LF}{bh^2} \tag{10.7}$$

For a circular section:

$$\sigma_t = -\frac{16LF}{3\pi D^3} \tag{10.8}$$

In these equations, F is the force at failure moment; L is the distance between the two supports; b and h are width and height of the specimen in the rectangular section, respectively; and D is the diameter of the specimen in the circular section. Although the four-point method is more accurate than the three-point method, the results obtained are not indicative of the direct tensile strength of rock and are usually 2–3 times of it. The values obtained from these methods are known as the failure modulus.

10.3.3 Hydraulic tension test

In this method, a specimen is prepared in the form of a ring with a thick wall, or a hollow cylinder, and then it is subjected to internal hydraulic pressure. In the method presented by Protodyakonov (1961), the specimen is subjected to internal hydrostatic pressure by a mixture of 75% resin and 25% paraffin.

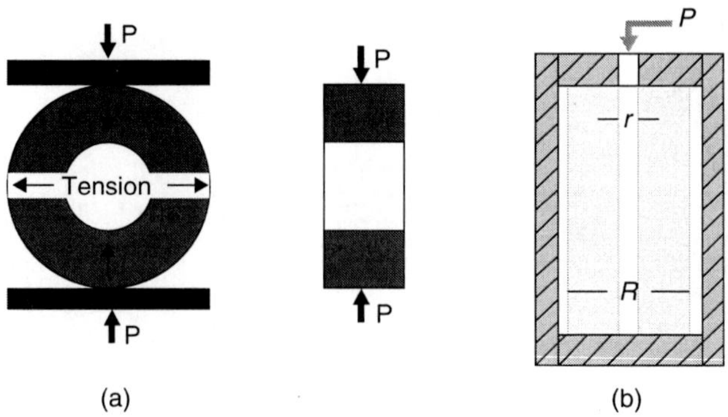

Figure 10.7 (a) Performing method of the ring tensile test and (b) its schematic view (Perras and Diederichs 2014).

10.3.4 Ring tensile test

In the ring tensile test, disc-shaped specimens with a hole in their center are used (Figure 10.7). This test is similar to the Brazilian test, and the maximum tensile strength value is calculated from the following equation:

$$\sigma_t = -\frac{2P\left(6+38r^2\right)}{\pi Dt} \tag{10.9}$$

In this equation, σ_t is the tensile strength, P is the load at the failure moment, r is the ratio of the inner radius to the outer radius $\left(\frac{r_i}{r_o}\right)$ of specimen, D is the diameter and t is the thickness of the specimen. There are many discussions about the radius of the specimen and the radius of the inner hole. Researchers have shown that the tensile strength decreases with increasing diameter of the inner hole and thickness of the specimen and increases with increasing specimen diameter. Hudson (1969) investigated the influence of the ratio of internal diameter to external diameter on tensile strength by conducting experiments on plaster with and without lime particles.

10.3.5 Cylinder diametric pressure test

In this test, for which various methods are also provided for performing it, cylindrical specimens are loaded in such a way that tension is created in them, and they break under this tension.

10.3.6 Sphere diametric pressure test

As a result of applying a radial compressive load to a spherical specimen, tensile failure occurs along a plane or planes passing through the loading diameter. For this purpose, the spherical specimen is placed inside the loading jaw and a force is applied to it in the direction of the diameter. The tensile strength in such a case is calculated from the following equation:

$$\sigma_t = -\frac{0.225P}{r^2} \tag{10.10}$$

In this equation, σ_t is the tensile strength, P is the load at the failure moment and r is the radius of the sphere.

10.3.7 Diametric pressure test of square plates

When a square plate is compressed by a pair of small flat jaws located on opposite sides, tensile stress develops in its central region. The loading mechanism in this method is shown in Figure 10.8.

10.3.8 Centrifuge tension test

The tensile strength of a rock specimen can be determined through centrifugal tension of the specimen. This work is done by a special centrifuge apparatus. The maximum rotation speed of this apparatus is such that it can produce a tension force of 10 MPa in the rock specimen.

10.3.9 Brazilian tensile strength test

The Brazilian tensile strength test is indirectly used to determine the tensile strength of intact rock cylindrical specimen. The justification of this test is based on the experimental fact that by applying the diametric pressure by loading jaws to a disc-shaped rock specimen, tensile stress perpendicularly spreads along the loading axis, and when this tensile stress exceeds the rock tensile strength, the specimen breaks (Figure 10.9). It should be noted that this test is not accurate enough for highly porous specimens, coarse-grained rocks, and specimens with joints and cracks.

The appropriate apparatus for applying and measuring the compressive load must have sufficient capacity and be able to apply the compressive load at the desired speeds. The loading

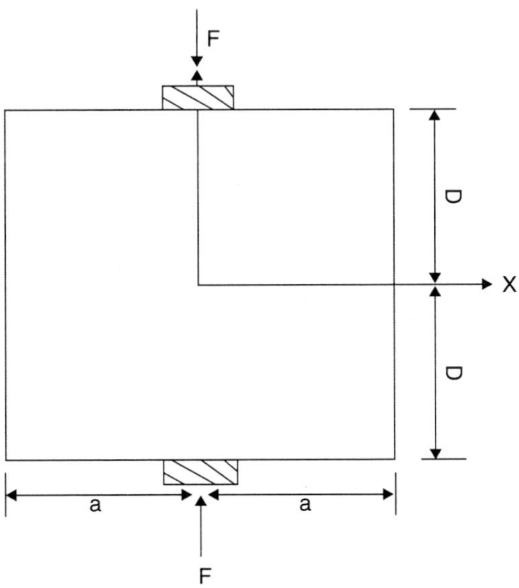

Figure 10.8 A schematic image of loading mechanism in the diametric pressure test of square plates.

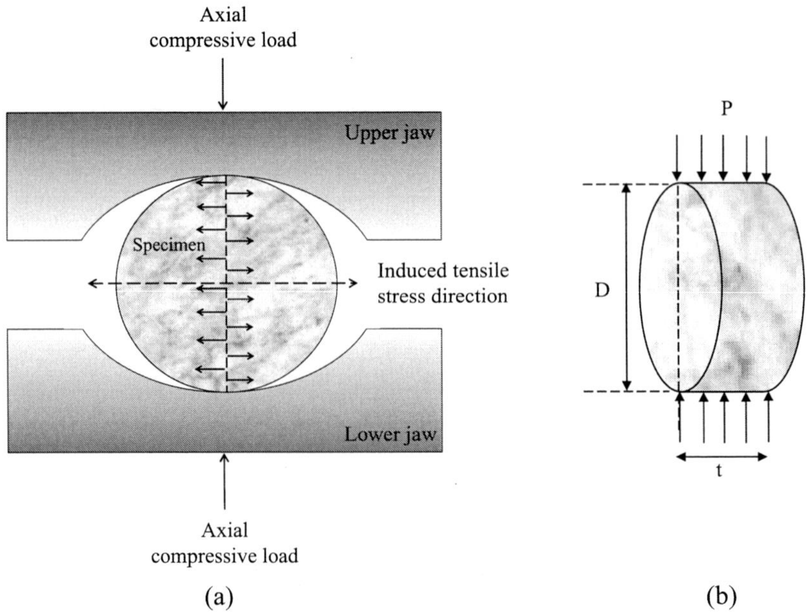

Figure 10.9 Performing mechanism of the Brazilian tensile strength test: (a) 2D view of specimen and test jaws and (b) 3D view of the specimen.

apparatus must be calibrated at appropriate time periods and conform to the conditions accepted in the ASTM (2008) standard. To place the loading plates of the apparatus parallel to each other, a spherical seat and ball are installed on the upper jaw of the apparatus. It is better for the loading apparatus to be equipped with a data recording system to draw the load-displacement curve. This helps to measure the breaking load. The tested specimens are cores with a diameter of at least 54 mm, whose head and bottom have been cut by a rock cutter (Jamshidi and Fereidooni 2022). The thickness (height) of the specimens should be approximately equal to their radius. The side surface of the specimens should be free of drilling marks, and their width and height should not exceed 0.025 mm. The end surfaces of the specimens should be smooth with an accuracy of 0.25 mm and be parallel to each other, while being perpendicular to the axis of the specimen with an accuracy of 0.25°. The orientation of the specimen with respect to the levels of anisotropy should be determined, and the moisture percentage of the specimens should be measured according to the existing standard methods. Also, the number of specimens is determined according to operational considerations, and usually ten specimens are suggested. The diameter and thickness of the specimen are determined with an accuracy of 0.25 mm and by averaging at least three measurements. Then, a layer of paper adhesive tape is attached around the specimen and diametrically marked on the specimen to determine the loading direction. To perform the test, a paper tape with a thickness of 0.2–0.4 mm and a width slightly greater than the thickness of the specimen is needed to wrap around it. The specimen is placed between the loading jaws in such a way that the load enters it diametrically. The arcuate surfaces should sit on the lateral surface of the specimen with no gaps between them. Then, the load is applied to the specimen continuously and at a uniform rate, so that the weakest rocks break within 15–30 seconds. A loading rate of 200 N/s is suggested. If the loading apparatus is equipped with a load-deformation recording system, the load corresponding to the occurrence of the first failure in the specimen must be accurately recorded, because in some cases, due to the separated pieces have load-bearing capacity even after the first failure, the specimen continues to bear the load. If the apparatus is not equipped with a data recording device, an attempt should be made to read the load at the first failure moment. At the moment of the first failure, there is a short interruption in the movement of the load gauge needle. However, the difference between the amount of load at the moment of initial failure, and its value at the final strength of the specimen is a maximum of 5%. The tensile strength obtained from the Brazilian test is most compatible with the direct tensile strength of rocks.

The performing method of the Brazilian tensile strength test is presented in ISRM (1978; 2007), ASTM (2008) and GOST (1985) standards. These standards have differences in some cases. But considering that in the method proposed by ISRM (1978; 2007), the stress distribution in the specimen is more uniform, and the results obtained from it are more reliable. In addition, today this method is more common. The comparison of the Brazilian test parameters and requirements according to different standards is summarized in Table 10.2. In this method, the indirect tensile strength can be calculated from the following relationship (Fereidooni and Karimi 2023):

$$\mathrm{BTS} = -\frac{2P}{\pi D t} = -0.636\,\frac{P}{Dt} \qquad (10.11)$$

In this equation, BTS is the Brazilian tensile strength, P is the load on the specimen, D is the specimen diameter and t is its thickness.

When arc-shaped loading jaws are used to perform the Brazilian tensile strength test, the steel jaws must be designed in such a way that they can grip a disc-shaped rock specimen diametrically so that they are tangential to it when it breaks in an arc of 10°. Critical dimensions in the

Table 10.2 Comparison of the Brazilian tensile strength test parameters and requirements as suggested by different standards

Parameter	*ISRM (1978; 2007)*	*ASTM (2008)*	*GOST (1985)*
Specimen diameter (mm)	54	42.2	40–50
Thickness to diameter ratio (*L/D*)	0.5	0.5–0.75	0.1
Measuring accuracy of specimen dimensions (mm)	–	0.25	0.1
Maximum roughness of lateral surface (mm)	0.025	0.5	0.4
Maximum roughness of end surfaces	0.25	–	0.4
Deviation from vertical end surfaces (deg.)	0.25	0.5	1
Loading jaws	Arc	Flat or arc	Flat with loading strip
Test duration (s)	15–30	10	–
Loading rate	200 N/s	300–500 PSI/min	0.1–1 MPa/s
Number of specimens	10	10	10
Calculating tensile strength formula	$\sigma_t = 0.636\frac{P}{Dt}$	$\sigma_t = \frac{2P}{\pi D t}$	$\sigma_t = 0.64\frac{P}{Dt}$

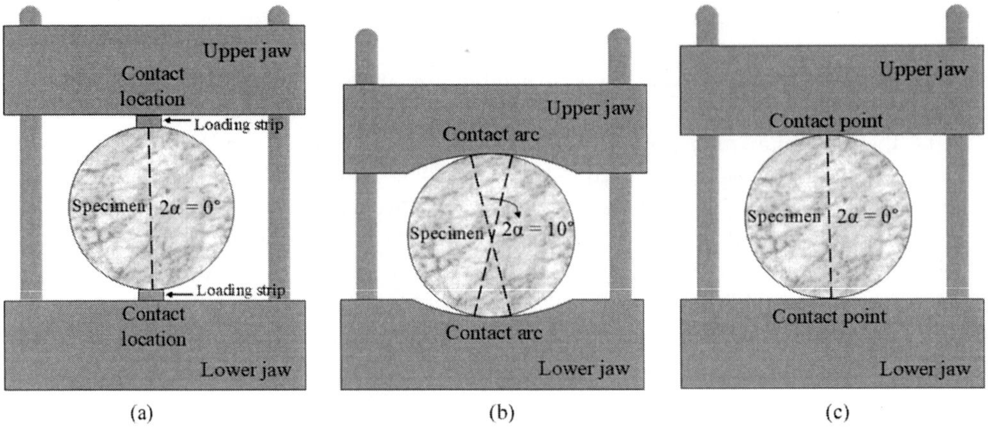

Figure 10.10 Comparison of jaws used in different Brazilian tensile strength test standards: (a) ASTM standard (2008), (b) ISRM standard (1978; 2007) and (c) GOST standard (1985) (Yousefi and Fereidooni 2020).

Table 10.3 Comparison of the Brazilian tensile strength test results with arched jaws and flat jaws

Rock type	*Brazilian tensile strength (MPa)*		
	Arched jaw	*Flat jaw*	*Difference*
Mudstone	4.9	4.2	0.7
Marl	5.8	5.2	0.6
Sandstone	10.8	10.4	0.4
Granite	14.3	13.9	0.4
Limestone	6.1	5.5	0.6

Brazilian testing machine include the radius of curvature of the jaws, the free distance between the pin and hole, the length of the pins connecting the two jaws and the width of the jaws. The radius of the jaws should be 1.5 times the radius of the specimen. The free distance between the pin and hole should be such that one jaw cannot rotate more than 4×10^3 Rad. relative to the other. In other words, for every 25 mm penetration of the pin into the jaw, the free distance should be 0.1 mm. Also, the width of the jaws is 1.1 times the thickness of the specimen. In Figure 10.10, the jaws proposed by the three standards of ASTM (2008), ISRM (1978; 2007) and GOST (1985) are schematically compared. Also, in Table 10.3, the results of the Brazilian tensile strength test with arched jaws and flat jaws are presented.

Yousefi and Fereidooni (2020) investigated the effect of jaw curvature, which is also known as the contact angle, and the central angle of the jaw arch (2α), on the results of the Brazilian indirect tensile strength of different rocks and achieved valuable results. Figure 10.11 shows failure patterns of a limestone specimen at different contact angles (2α). They found that Brazilian tensile strength at different 2α angles (i.e. 2α equal to zero, 10°, 15°, 20°, 45° and 60°) was different for the studied rocks (Figure 10.12). This means that the Brazilian tensile strength increases with increasing 2α angle from 0° to 60°. Also, it was found that in most of the studied rocks, the highest change rate in the Brazilian tensile strength occurs in the range of the

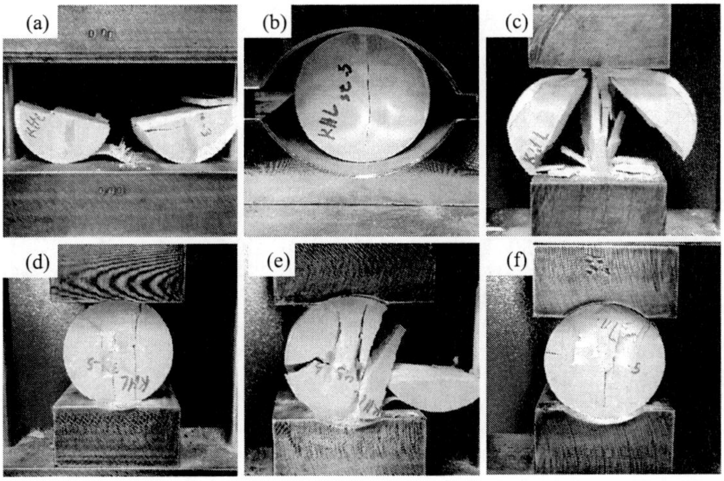

Figure 10.11 Limestone failure patterns at different contact angles ($2a$): (a) 0°, (b) 10°, (c) 15°, (d) 30°, (d) 45° and (c) 60° (Yousefi and Fereidooni 2020).

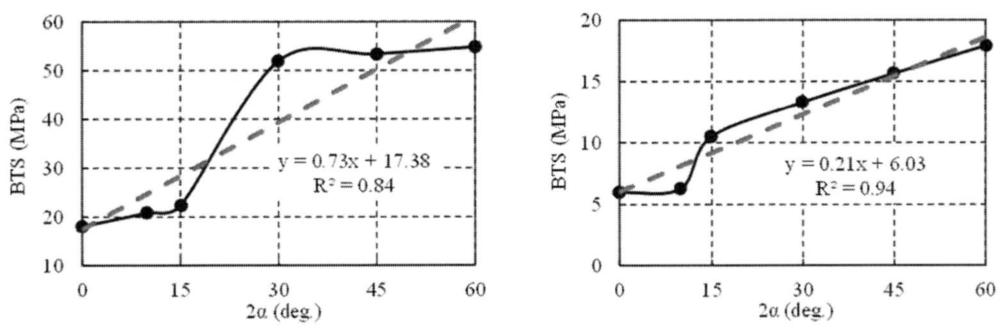

Figure 10.12 Relationship between the Brazilian tensile strength (BTS) and contact angle ($2a$) in a type of (a) limestone and (b) skarn (Yousefi and Fereidooni 2020).

mentioned angle between 15° and 30°. On the other hand, the difference in Brazilian tensile strength at angles from 0° to 10° has the least change, and this means that the jaws proposed by different standards do not differ much from each other in determining tensile strength.

The research results of Fereidooni (2014) and Fereidooni et al. (2016) show that in anisotropic rocks, the amount of Brazilian tensile strength is strongly influenced by the orientation of rock anisotropy planes. So that in regional metamorphic rock specimens such as slate, phyllite and schist, the minimum value of Brazilian tensile strength is obtained at anisotropy angles of 15° or 30° with respect to the loading direction of the specimens. While in contact metamorphic rocks such as hornfels, the minimum Brazilian tensile strength is obtained at an anisotropy angle of 0°. In all anisotropic metamorphic rocks, the maximum Brazilian tensile strength is obtained at an anisotropy angle of 90° with respect to the loading direction of specimen.

At the end of this chapter, the range of mechanical properties of some common rocks is presented in Table 10.4 in order to know and compare these values with each other. In this regard, igneous rocks are usually more resistant and less deformable than sedimentary, and metamorphic ones.

Table 10.4 Value range of mechanical properties of some common rocks

Rock type		*Uniaxial compressive strength (MPa)*	*Elasticity modulus (GPa)*	*Poisson's ratio*	*Point load strength (MPa)*	*Tensile strength (MPa)*	*Failure strain (%)*
Igneous	Granite	100–300	30–70	0.17	5–15	7–25	0.25
	Diorite	100–350	30–100	0.10–0.20		7–30	0.30
	Gabbro	150–250	40–100	0.2–0.35	6–15	7–30	0.30
	Rhyolite	80–160	10–50	0.20–0.40		5–10	
	Andesite	100–300	10–70	0.20	10–15	5–15	
	Basalt	100–350	40–80	0.10–0.20	9–15	10–30	0.35
Sedimentary	Conglomerate	30–230	10–90	0.10–0.15		3–10	0.16
	Sandstone	20–170	15–50	0.14	1–8	4–25	0.20
	Shale	5–100	5–30	0.10		2–10	
	Mudstone	10–100	5–70	0.15	0.10–6	5–30	0.15
	Dolomite	20–120	30–70	0.15		6–15	0.17
	Limestone	30–250	20–70	0.3	3–7	6–25	
Metamorphic	Gneiss	100–250	30–80	0.24	5–15	7–20	0.12
	Schist	70–150	5–60	0.15–0.25	5–10	4–10	
	Phyllite	5–150	10–85	0.26		6–20	
	Slate	50–180	20–90	0.20–0.30	1–9	7–20	0.35
	Marble	50–200	30–70	0.15–0.30	4–12	7–20	0.40
	Quartzite	150–300	50–90	0.17	5–15	5–20	0.20

References

ASTM D 3967-08, 2008. *Standard Test Method for Splitting Tensile Strength of Intact Rock Core Specimens*. ASTM International, West Conshohocken, PA.

ASTM D2936-20, 2020. *Standard Test Method for Direct Tensile Strength of Intact Rock Core Specimens*. ASTM International, West Conshohocken, PA.

Brace, W.F., 1964. Brittle fracture of rocks. In: Judd, W.R. (ed) *Proceedings of the International Conference on State of Stress in the Earth's Crust* (111–174). Elsevier, New York.

Broch, E., Franklin, J.A., 1972. The point-load strength test. *International Journal of Rock Mechanics and Mining Sciences & Geomechanics Abstracts*, 9(6), 669–676.

Fairhurst, C., 1961. Laboratory measurements of some physical properties of rock. *Proceedings of the Fourth Symposium on Rock Mechanics*, Pennsylvania, USA.

Fereidooni, D., Karimi, Z., 2023. Predicting rock brittleness indices from simple laboratory test results using some machine learning methods. *Geomechanics and Engineering*, 34(6), 697–726.

Fereidooni, D., Khajevand, R., 2018. Determining the geotechnical characteristics of some sedimentary rocks from Iran with an emphasis on the correlations between physical, index, and mechanical properties. *Geotechnical Testing Journal*, 41(3), 555–573.

Fereidooni, D., 2004. Investigating the stability of rock slopes along the Ganjnameh-Shahrestaneh road. Master's Thesis in Engineering Geology, Faculty of Science, Bu-Ali Sina University, 192 pp.

Fereidooni, D., 2014. The study of engineering geological properties of metamorphic rocks in Hamedan province with emphasis on anisotropy. PhD Thesis in Engineering Geology, Bu-Ali Sina University, Hamedan, Iran, 224 pp.

Fereidooni, D., 2016. Determination of the geotechnical characteristics of hornfelsic rocks with a particular emphasis on the correlation between physical and mechanical properties. *Rock Mechanics and Rock Engineering*, 49, 2595–2608.

Fereidooni, D., Khanlari, G.R., Heidari, M., Sepahigero, A.A., Kolahi-Azar, A.P., 2016. Assessment of inherent anisotropy and confining pressure influences on mechanical behavior of anisotropic foliated rocks under triaxial compression. *Rock Mechanics and Rock Engineering*, 49, 2155–2163.

Gorski, B.P., 1993. Tensile testing apparatus. U.S. Patent 5193396.

Hoek, E., 1964. Fracture of anisotropic rock. *Journal of the Southern African Institute of Mining and Metallurgy*, 64(10), 501–518.

Hudson, J.A., 1969. Tensile strength and the ring test. *International Journal of Rock Mechanics and Mining Sciences & Geomechanics Abstract*, 6(1), 91–97.

ISRM, 1978. Suggested methods for determining tensile strength of rock materials. *International Journal of Rock Mechanics and Mining Sciences & Geomechanics Abstracts*, 15(3), 99–103.

ISRM, 2007. Compilation arranged by the ISRM Turkish National Group, Ankara, Turkey. In: Ulusay, R., Hudson, J.A. (eds) *The Blue Book: The Complete ISRM Suggested Methods for Rock Characterization, Testing and Monitoring, 1974–2006*. Kazan Offset Press, Ankara, Turkey.

Jamshidi, A., Fereidooni, D., 2022. Evaluation of the block punch index test for predicting the strength of sandstones. *Scientific Quarterly Journal of Iranian Association of Engineering Geology*, 15(1), 105–118.

Khanlari, G.R., Heidari, M., Sepahigero, A.A., Fereidooni, D., 2014. Quantification of strength anisotropy of metamorphic rocks of the Hamedan province, Iran, as determined from cylindrical punch, point load and Brazilian tests. *Engineering Geology*, 169, 80–90.

Klanphumeesri, S., 2010. Direct tension testing of rock specimens. Doctoral dissertation, School of Geotechnology Institute of Engineering Suranaree University of Technology.

Margan, B., Fereidooni, D., 2024. Evaluation of rock burst phenomenon in tunnel sites using experimental and semi-experimental methods – A case study: Haji-Abad tunnel in Hormozgan Province. *Journal of Engineering Geology*, 18(1), 71–88, (in Persian).

Obert, L., Windes, S.L., Duvall, W.I., 1946. Standardized tests for determining the physical properties of mine rock. U.S. Bur. Mines Rept. Invest., No. 3891.

Perras, M.A., Diederichs, M.S., 2014. A review of the tensile strength of rock: concepts and testing. *Geotechnical and Geological Engineering*, 32, 525–546.

Protodyakonov, M.M., 1961. Methods of Studying the Strength of Rocks used in the U.S.S.R. *Proceedings of the International Symposium on Mining Research*, University of Missouri.

Shariati, M., Fereidooni, D., 2021. Rock slope stability evaluation using kinematic and kinetic methods along the Kamyaran-Marivan road, west of Iran. *Journal of Mountain Science,* 18(3), 779–793.

Townhill-Rewston, P., 1984. In towards a relational scale of hardness for minerals. In: Dowd, P.A. (ed) *Relational Scale of Hardness for Minerals* (22–27). Leeds University Mining Association, West Yorkshire.

USSR State Standard GOST 21153.3-85, Rocks, 1985. *Methods for Determining Ultimate Strength under Uniaxial Tension*. Publisher of Standards, Moscow.

Wuerker, R.G., 1955. Measuring the tensile strength of rocks. *Minerals Engineering International*, 7(2), 157–168.

Yousefi, H., Fereidooni, D., 2020. The effect of jaw's curvature on Brazilian tensile strength of rocks. *Geomechanics and Engineering*, 23(2), 165–178.

Chapter 11

Shear strength

11.1 Introduction

Shear strength is the stability and resistance of rock and especially its discontinuity surfaces against shear forces. The shear forces are a pair of forces that are equal, parallel, but in opposite directions and act on the specimen and cause a shear failure surface in it. The triaxial compressive strength test is used in the laboratory to determine the shear strength of intact rock and its parameters. To determine the shear strength parameters of discontinuity surfaces of rock mass, the direct shear test is suitable and used on artificial (saw cut) or natural rock joint surfaces. It is obvious that the shear strength parameters of intact rock are always higher than the shear strength parameters of discontinuities. The discontinuities are weak planes in rock mass, which show much less strength against external forces such as shear forces, especially when the discontinuity is open and has lost its cohesion. Therefore, it is very important to identify discontinuity type, amount of opening and its filling materials for determining the shear strength of rock mass discontinuities (Fereidooni 2018; Fereidooni et al. 2015). In this chapter, the shear strength issue of intact rock and rock mass discontinuity surfaces is examined in detail.

11.2 Triaxial shear strength

According to the methods suggested by ISRM (1983; 2007) and ASTM-D-2664 (1995), the triaxial shear strength test should be performed at least in five different lateral (confining) pressures (σ_3) (for example, 1, 3, 5, 7 and 9 MPa) on five cylindrical rock specimens and the maximum axial stress value (σ_1) corresponding to each lateral pressure is measured. Then, with this information, to determine the shear strength parameters, namely, cohesion (c) and internal friction angle (ϕ) of intact rock, the following two methods can be used:

11.2.1 Drawing Mohr's circles and Mohr–Coulomb failure envelope

In this method, Mohr's circles regarding the tests are first drawn. For each test, these circles are drawn to the center of $\frac{\sigma_1+\sigma_3}{2}$ and to the radius of $\frac{\sigma_1-\sigma_3}{2}$ in a shear stress-normal stress coordinate system, and then the best possible line, which is called the failure envelope, touches these circles (Figure 11.1). If the envelope is in a straight line, the angle of this line with the horizon is the internal friction angle (ϕ), and its width from the origin is the rock cohesion (c). If the failure envelope is non-linear, to determine the cohesion and internal friction angle, it is necessary to make a line tangent to the envelope at the contact point with each Mohr circle and

DOI: 10.1201/9781003672920-11

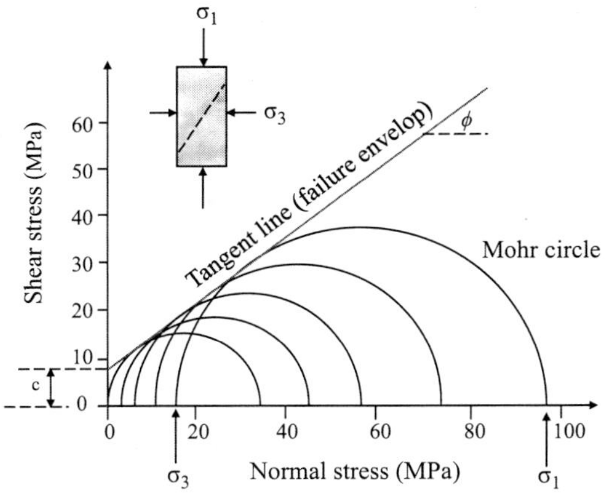

Figure 11.1 Mohr's circle drawing based on the triaxial compressive strength test to calculate the internal friction angle and cohesion of intact rock.

consider the slope angle and width from the origin of the tangent line as ϕ and c. In this way, we will have a separate cohesion and internal friction angle for each lateral pressure. In other words, in such cases, ϕ and c depend on the level of normal or lateral stresses (Diamantis and Fereidooni 2023). In this case, the value of the internal friction angle can be calculated from the following equation:

$$\phi = \text{arc tan}\left(\frac{\Delta\tau}{\Delta\sigma_n}\right) \tag{11.1}$$

In this equation, $\Delta\tau$ and $\Delta\sigma_n$ are, respectively, the changes of shear and normal stresses in the range where the internal friction angle is measured.

11.2.2 Drawing the diagram of the main stresses

In this method, the lateral pressure and strength values corresponding to each of the lateral pressures in the diagram of the main stresses ($\sigma_1 - \sigma_3$) are established as separate points, and the resistance envelope is obtained by fitting the best curve to these points (Figure 11.2). This curve may be linear or it can be divided into several linear parts. In this way, the slope (m) and the width from the origin (b) of each linear part are known. By having the parameters m and b, according to Columb's failure theory, the values of ϕ and c can be determined from the following relationships (Fereidooni and Khajevand 2019):

$$\phi = \text{arc sin}\left(\frac{m-1}{m+1}\right) \tag{11.2}$$

$$c = b\left(\frac{1-\sin\phi}{2\cos\phi}\right) \tag{11.3}$$

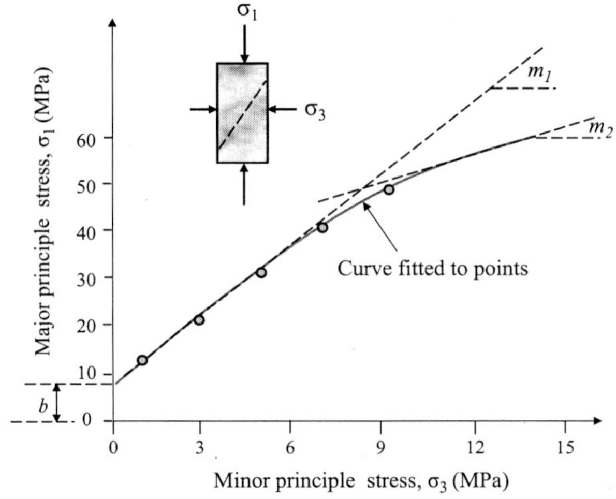

Figure 11.2 Drawing of the main diagram of stresses based on the triaxial compressive strength test to calculate the internal friction angle and cohesion of intact rock.

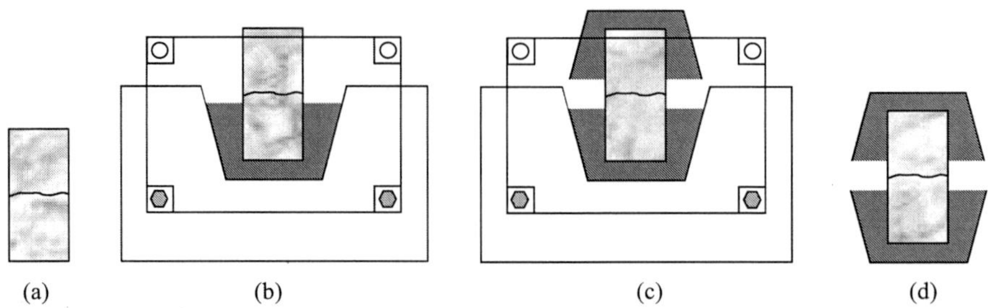

Figure 11.3 Preparation steps of a cast core specimen with a natural joint to perform the direct shear test: (a) natural joint in a rock core, (b) casting an end of the rock core, (c) casting the other end of the rock core and (d) specimen is ready to be placed in the direct shear test apparatus.

11.3 Joint surface shear strength

Shear strength of joint surfaces and rock mass discontinuities is very important in stability analysis of rock structures such as rock slopes. Achieving appropriate shear strength values requires a correct understanding of the theoretical foundations of shear failure and a complete knowledge of rock mass joint and discontinuity conditions (Fereidooni 2015; Shariati and Fereidooni 2021). In the laboratory, to calculate the shear strength parameters of the joints, the direct shear test is performed on artificial (saw cut) or natural joint surfaces according to ISRM (2007), ASTM-D-5607 (2016) and JGS (2008) standards. To prepare a specimen with a saw cut or artificial joint surface, after preparing a cylindrical rock specimen, it should be cut by a cutting machine perpendicular to the cylinder axis. To prepare a specimen with a natural joint surface, the rock specimen should be cored in the direction perpendicular to the joint plane. Then, the two ends of the specimen with a saw cut or natural joint surface are cast by special plaster (Fereidooni 2004). Figure 11.3 shows the steps of casting a cylindrical rock specimen

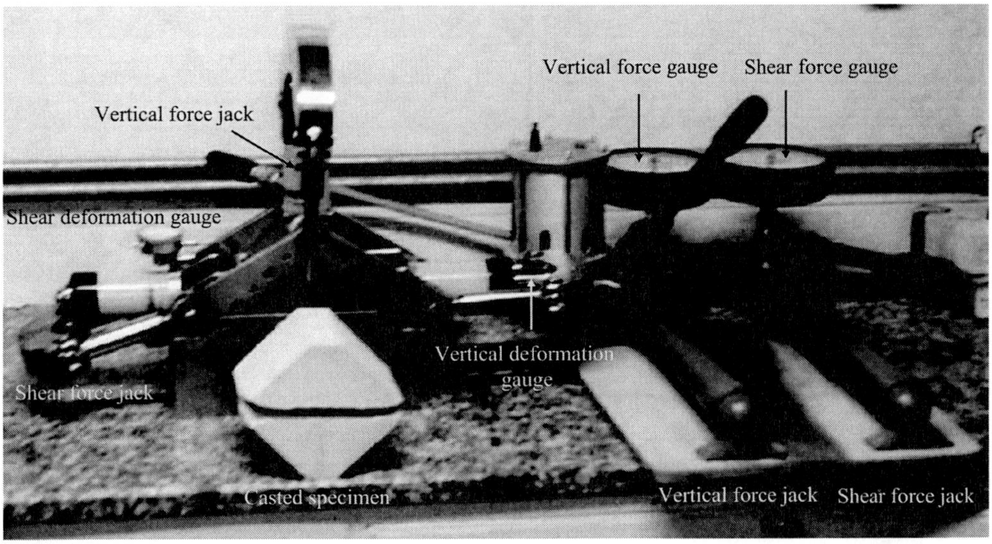

Figure 11.4 Rock direct shear test apparatus and its different parts (Fereidooni 2004).

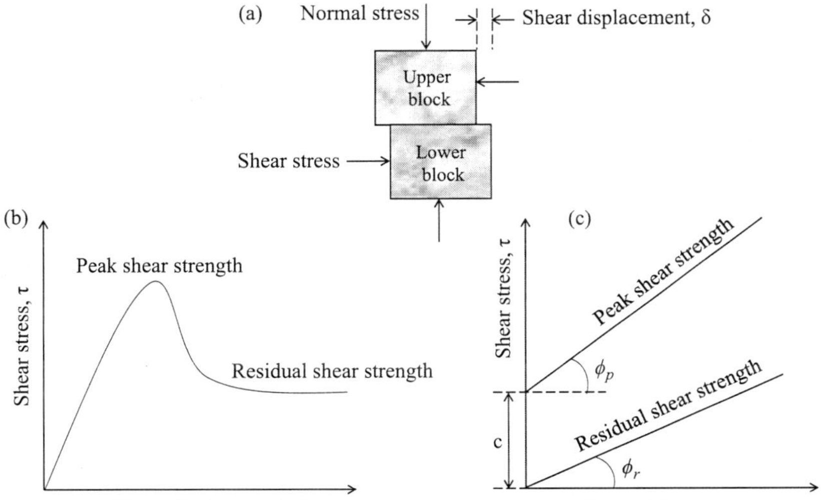

Figure 11.5 Curves resulting from the direct shear test on smooth or saw cut joint surfaces: (a) direct shear test mechanism, (b) shear stress-shear displacement curve and (c) shear stress-normal stress curve.

to perform a direct shear test on the joint surface, and Figure 11.4 shows an example of a rock direct shear strength test apparatus and its different parts.

The result of the direct shear test is the drawing of two curves of shear stress-shear displacement and shear stress-normal stress. Figure 11.5 shows the mechanism of direct shear test and the resulting curves for artificial smooth joint surfaces. As can be seen in Figure 11.5b, after a relatively linear deformation, the specimen reaches the peak (ultimate) shear strength and continues to move after failure in a constant shear stress that is less than the peak shear stress. This constant stress is called residual shear strength.

If several direct shear tests are performed at different normal stresses and their peak and residual shear strength values are established in a coordinate system such that the vertical axis is the shear stress and the horizontal axis is the vertical stress, the curves presented in Figure 11.5c are obtained. One of the two curves with a higher slope is related to the peak shear strength and the curve with a lower slope belongs to the residual shear stress. The slope of these curves represents the peak and residual friction angle (ϕ_p and ϕ_r), and their width from the origin represents the cohesive joint surface (c). As can be seen, the width from the origin of the residual shear resistance curve is very small and is considered equal to zero. The equations of these two curves are as follows.

Peak shear strength:

$$\tau_p = c + \sigma_n \tan\phi_p \tag{11.4}$$

Residual shear strength:

$$\tau_r = \sigma_n \tan\phi_r \tag{11.5}$$

The surfaces of natural discontinuities in hard rocks are never as smooth as the surfaces of saw cut or polished joints. This roughness increases the shear strength of the joint surfaces. The effect of joint roughness can be studied and analyzed using Patton's criterion (1966). According to this criterion, the following relationship is presented to determine the shear strength of tested specimens with rough surfaces:

$$\tau = \sigma_n \tan\left(\phi_b + i\right) \tag{11.6}$$

In this equation, ϕ_b is the basic friction angle of joint surface and i is the inclination angle of the joint surface. Figure 11.6 shows the concept of Patton's criterion (1966) to check the shear strength of joints with a serrated surface. Therefore, a sloping or serrated discontinuity will have an apparent friction angle equal to ($\phi_b + i$). Patton (1966) found out by experimenting on specimens with different shapes and compositions that at low normal stresses, sliding occurs

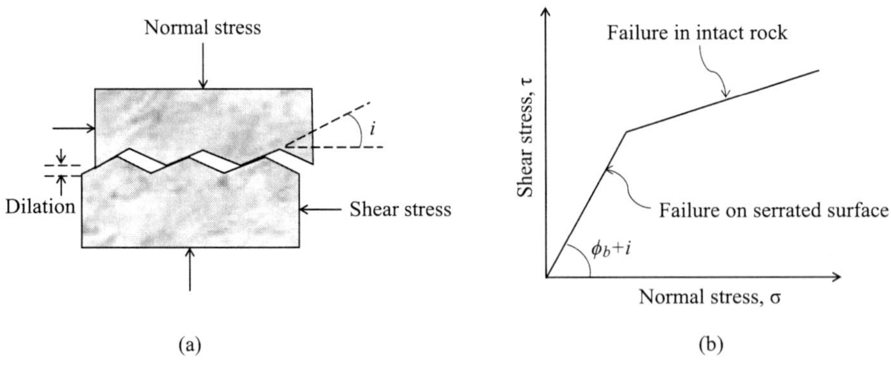

Figure 11.6 Patton's criterion (1966) regarding the shear strength of specimens with a serrated surface: (a) inclined or serrated discontinuity which is dilated and (b) bilinear graph of shear strength (Fereidooni 2023).

on the discontinuity surface based on equation (11.6), and in this case, we will face the dilation (Figure 11.6a). As the value of normal stress rises above a critical limit, sliding on the serrated inclined surfaces no longer occurs, and shear stress causes teeth to be cut. In this way, the shear strength diagram for such specimens is bilinear as shown in Figure 11.6b. With the increase of normal stress, the amount of dilation decreases, and finally, we reach a state where cutting occurs on intact rock material.

Barton and Bandis (1990) presented another failure criterion for predicting the shear strength of rough joints, based on the precise tests they conducted on artificial rough joints, which is in the form of the following relationship:

$$\tau = \sigma_n \tan\left[\phi_r + \text{JRC}.\log\left(\frac{\text{JCS}}{\sigma_n}\right)\right] \tag{11.7}$$

In this equation, JRC is the joint roughness coefficient, JCS is the compressive strength of the joint wall, ϕ_r is the permanent or residual friction angle and σ_n is normal stress. This equation is best used in the range of $0.01 < \frac{\sigma_n}{\text{JCS}} < 0.3$. It is worth mentioning that as the ratio of $\frac{\sigma_n}{\text{JCS}}$ approaches zero in Barton's equation, the value of $\frac{\text{JCS}}{\sigma_n}$ tends to infinity and this equation becomes invalid. In this equation, the value inside the bracket should not be more than 70°.

The tilt test or the joint surface profile gauge can be used to determine the joint roughness coefficient (JRC). Figure 11.7 shows how to determine the joint surface roughness by a profile gauge. For this purpose, first the profile gauge is placed on the joint surface and profile obtained from the joint surface is drawn on a millimeter paper (Figure 11.7a). Then, the visually drawn profile is compared with the ten standard profiles of the joint surface measurement scale shown in Figure 11.7b, and whichever has conditions closer to the actual roughness is selected as the joint surface roughness coefficient.

The Schmidt hammer is used to determine the compressive strength of the joint wall (JCS), which was explained in detail in the hardness and abrasiveness chapter. It should be noted that

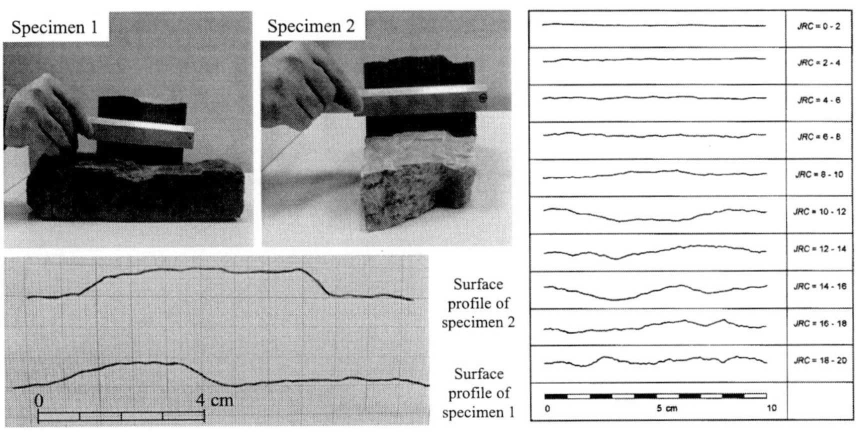

Figure 11.7 (a) Measuring method of the joint surface roughness and (b) standard scale for measuring the joint roughness coefficient.

the Schmidt hammer test provides a measure of the mechanical strength of the thin layer of weathered material adjacent to the failure surface. Considering that these weathered materials together with the surface roughness of the discontinuity control its shear strength, it is considered an important indicator in this field (Barton and Bandis 1990). When determining the Schmidt hardness and dry density of rock, the compressive strength of the joint wall will be calculated using experimental relationships (refer to Fereidooni 2023). The most important factors affecting the rock shear strength are weathering, shear surface roughness and the presence of water under pressure, and other differences may exist in the laboratory and in situ scales.

11.4 Tilt test

Patton (1966) and Goldstein et al. (1966) were probably the first authors to study the peak shear strength of rock discontinuities and provided the principles upon which various approaches were developed. In this regard, Barton (1973), Barton and Bandis (1982) and Barton and Choubey (1977) presented relations for estimating the shear strength of rough, unfilled and unweathered rock joints. For this purpose, they used a simple test in which they placed two pieces of small rock on top of each other and slowly tilted them until the upper piece started to slide on the lower piece. This test was later called as the tilt test. However, the first case of a tilt test as a sliding method for estimating the frictional behavior of rock mass joints was presented by Hoek and Bray (1974), who showed that the shear strength of discontinuities can be estimated using the tilt test. This topic was included as part of the Cawsey (1974) doctoral thesis as the initial design of the tilt test and the test program. Then, Cawsey and Farrar (1976), after many efforts, presented a device to measure the joint shear strength. This device consisted of a set of two wooden plates that were connected on one side and the upper plate could move up and down. The device was also equipped with a scale to measure the slip angle. They recommended an accuracy of ±0.5° for the slip angle measurement. In addition, Cawsey and Farrar (1976) presented some preliminary results on the water effect on the frictional behavior of gypsum joint surfaces as well as the effect of removing rock dust from sliding surfaces in the tilt test. These studies included not only the improvement of the tilt test procedure, but also the interpretation of the test results and conditions.

For smooth or saw cut joint surfaces, based on the geometric shape of the upper and lower blocks and the contact surface of the two blocks, the tilt test according to Figure 11.8 can be performed in one of the following four modes (Alejano et al. 2018):

1 Rectangular cube blocks with a rectangular contact surface
2 Longitudinal cut core with a rectangular contact surface
3 Triple cores with linear contacts (known as Stimpson's method)
4 Double cores with linear contact

Barton and Choubey (1977) also used the tilt test to estimate the basic friction angle values (ϕ_b) of flat and saw cut surfaces of rock. According to their studies, when the tilt test is performed on a rough joint, the angle at which the slip occurs may be 40° or 50° more than the basic friction angle or even more than the residual friction angle (ϕ_r). Based on the reports published by Barton and Choubey (1977), the basic friction angle for smooth and unweathered rock surfaces is between 25° and 35°.

Stimpson (1981) proposed a different conceptual method for tilt tests of rough planar surfaces. He proposed a new type of the test using three rock cores obtained by common core drilling methods.

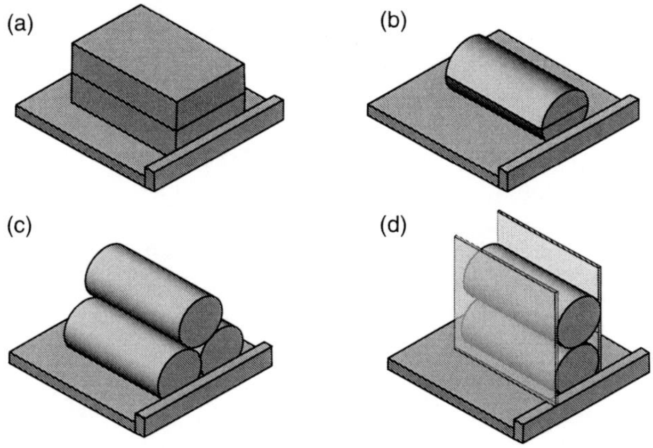

Figure 11.8 Different modes of the tilt test depending on contact surface type (Alejano et al. 2018): (a) surface contact of rectangular cube blocks, (b) contact of a core specimen with a longitudinal cut, (c) linear contacts in triple cores (known as Stimpson's method) and (d) linear contact in double cores.

For this purpose, three cores should be placed on a base that can be tilted slowly, and the angle at the moment of sliding should be recorded with an accuracy of ±0.5°. The most important advantage of this method was the possibility of testing rock cores instead of smooth surfaces. In Stimpson's proposed method, the following equation is used to obtain the basic friction angle of specific rock surfaces (ϕ_A), by measuring the slip angle (α) at which the top core slides:

$$\phi_A = \tan^{-1}\left(\frac{2}{\sqrt{3}}\tan\alpha\right) \tag{11.8}$$

This relationship showed an error that led to overestimation of the tilt test results over several years. After this method, an advanced and newer device to perform the tilt test was presented by Bruce et al. (1989). This device had new features such as an electric readout and was equipped with clamps to hold rock specimens with dimensions of 5 × 5 to 15 × 15 cm. In addition, this device was able to perform the tilt tests at different rotation rates. Although this method in itself did not improve the traditional methods, it allowed for the development of the first studies on the effect of test plate lifting velocity on the results. Then, Barton (2011) proposed a new arrangement for the tilt test, in which two cores were used instead of three rock cores suggested by Stimpson (1981) (Figure 11.9).

Alejano et al. (2012) presented a tilt test device. The device consisted of a metal frame with a flat plate attached to a rod and a gearbox driven by an asynchronous motor. The device was controlled by a two-switch controller that performed both raising and lowering movements of the flat plate (Figure 11.10a). The new device allowed the tilt test to be performed with greater comfort as well as a controlled lifting velocity. This device was an evolved version of the older hand-operated tilt tests (Figure 11.10b and c).

These researchers identified an error in the relationship presented by Stimpson (1981) and modified this relationship as follows:

$$\phi_A = \tan^{-1}\left(\frac{\sqrt{3}}{2}\tan\alpha\right) \tag{11.9}$$

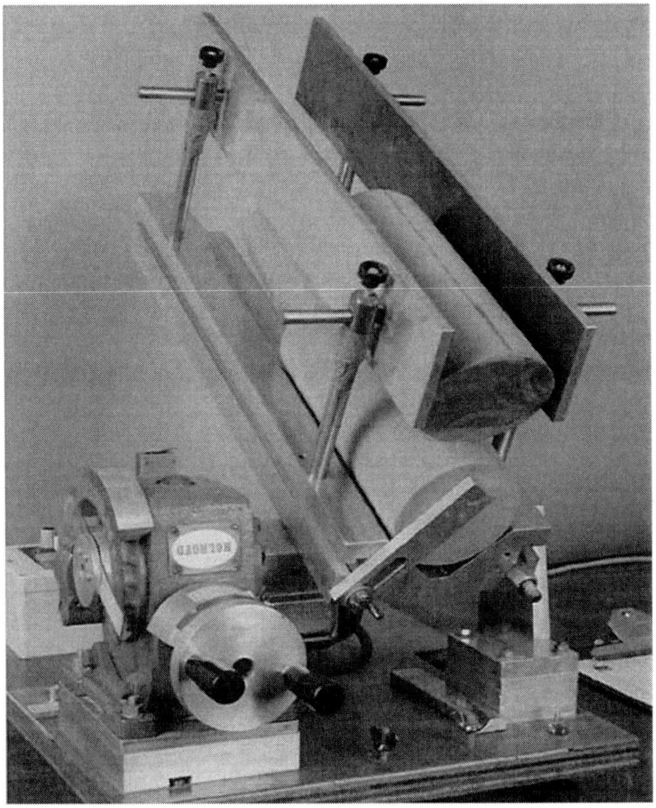

Figure 11.9 Double core tilt test (Barton 2011).

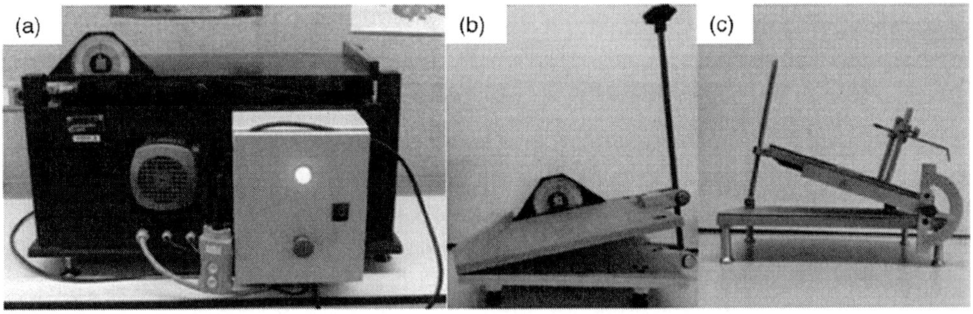

Figure 11.10 (a and b) Manual tilt test device and (c) tilt test device powered by an electric motor (Alejano et al. 2012).

Three-core tilt tests show unreliable friction angles of rock joints due to overestimation of results, and tests with surfaces less than 50 cm^2 are not recommended. Therefore, due to the easier theoretical analysis and ensuring the distribution of compressive stresses along the contact surfaces, square-shaped specimens with a length-to-thickness ratio greater than 2 have been suggested (Muralha 2012).

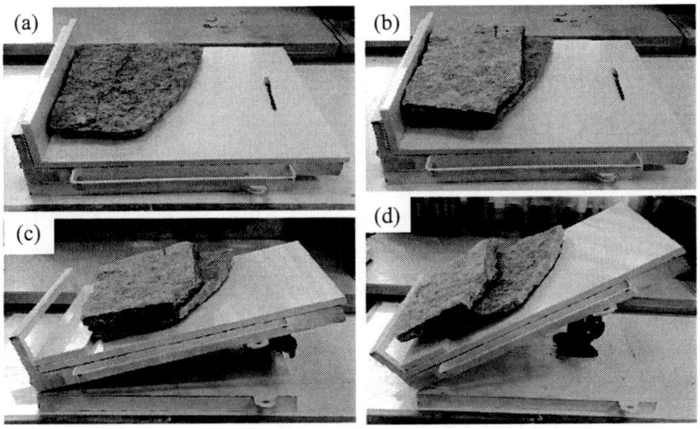

Figure 11.11 Performing mechanism of the tilt test: (a) the lower half of rock block is placed horizontally on the test plate, (b) the upper half of the block is placed on its lower half which is fixed, (c) the test plate is inclined and (d) the upper half of the block slides and the slip angle (α) is measured by a protractor.

Alejano et al. (2012) estimated the basic friction angle of rock joints using an electric tilt test. The performing steps of the tilt test in this research are as follows:

1. Specimens are cut with specific dimensions.
2. The lower half of rock block is placed on test plate. In this case, the test plate should be horizontal. For this purpose, a level bubble is used to check the horizontality of the test plate.
3. The upper half of the block is placed on its lower half, which is fixed.
4. The test plate is tilted at a rate of 0.4°/s until the upper half of the block slides, and in this case, the slope of the plate is recorded.
5. Since small displacements may sometimes occur during full displacement, only tests in which these displacements amount to at least 10% of the specimen length are considered valid.
6. Each test is repeated at least three times. Before repeating the test, the surfaces of the specimens are cleaned with a dry cloth or brush, and then the specimens are placed in the same initial position.
7. The results of all tests are averaged and used to calculate different parameters.

Kim et al. (2016) studied the effect of contact surface stress distribution on the results of the tilt test using a manual tilt test device. The innovation of this study was the use of truncated cylinder specimens to determine the basic friction angle. Ulusay and Karakul (2016) conducted a suitable number of tilt tests in dry, wet and submerged conditions to determine the dependence of the basic friction angle on different tilt test conditions and evaluated the exact effect of water height on the basic friction angle in submerged conditions.

Fathalizadeh (2022) investigated the friction angle of rock joint surfaces in different directions of a specimen surface and the upper block weight effect using a manual tilt test device made in Damghan University. In general, a suitable specimen is first prepared to perform the tilt test. According to the method suggested by ISRM (2007), the most suitable geometry for the tilt tests is to use sheet-shaped specimens with a length-to-height ratio greater than 4, to ensure proper distribution of compressive stress during sliding. Also, considering the contact surfaces of the two blocks, it is recommended that these surfaces be larger than 50 cm^2 and at least a width of 55 mm. The device used to perform the tilt test must be able to stop the process of tilting the test surface at the desired time, and have a suitable tool with a minimum accuracy of ±0.5° to measure the slip angle. After preparing the specimens, the lower part of the specimen is placed on the test plate. In this case, the test plate should be horizontal. For this purpose, a level bubble is used to check the horizontality of the test plate. Then, the upper part of the specimen is placed on the fixed lower part and the test plate is tilted by a jack until the upper part of the specimen slides. Since it is sometimes possible for small displacements to occur during whole displacement, only tests in which these displacements amount to at least 10% of the length of the specimen are considered valid. Each test is repeated at least three times, and before repeating the test, the surfaces of the specimens are cleaned with a dry cloth or brush, and then the specimens are placed in the same initial position. The performing mechanism of the tilt test proposed by Fathalizadeh (2022) are presented in Figure 11.11.

In the tilt test, the angle at which slip occurs (slip angle, α) is a function of the ratio between the shear stress (τ) and the effective normal stress (σ_{n0}). The effective normal stress created by the earth's gravity force that enters the upper half of the block, and its value at the unit area, can be calculated as follows:

$$\sigma_{n0} = \frac{W}{A}\cos\alpha = \gamma b \cos\alpha \tag{11.10}$$

In this equation, W is the weight of the upper half of the specimen (N), A is the contact surface area of the two halves of the specimen, γ is dry unit weight (N/m^3), b is thickness of the upper half of the specimen block and α is slip angle of the upper half of the specimen block. In this test, the value of the joint roughness coefficient obtained from the tilt test can also be calculated through the following equation:

$$\text{JRC} = \frac{\alpha - \varphi_r}{\log_{10}\left(\frac{\text{JCS}}{\sigma_{n0}}\right)} \tag{11.11}$$

In this equation, ϕ_r is the residual friction angle and JCS is the joint compressive strength. The tilt test is done only in dry joints to avoid any possible problems with joint water pressure fluctuations. Thus, the value of JCS is the same as the value measured on the dry joint by the Schmidt hammer. The values of damage coefficient (M) and peak dilation angle (d_n) of the joint surface in the tilt test can also be calculated from the following relations:

$$M = \frac{\text{JRC}}{12.\log\frac{\text{JCS}}{\sigma_{n0}}} + 0.70 \tag{11.12}$$

$$d_n = \frac{12\text{JRC}\left(\log\frac{\text{JCS}}{\sigma_{n0}}\right)^2}{\text{JRC} + 8.4\log\frac{\text{JCS}}{\sigma_{n0}}} \tag{11.13}$$

The following relationship is also used as a basis for estimating the maximum rock surface shear stiffness (K_s). The maximum shear stiffness is defined as the maximum shear strength divided by the maximum shear displacement. In fact, examination of in-situ and laboratory shear tests shows that the shear stiffness is inversely related to the joint length:

$$K_s = \frac{100}{L}\sigma_n \tan \mathrm{JRC} \log \frac{\mathrm{JCS}}{\sigma_n} + \varphi_r \tag{11.14}$$

Finally, the shear strength of rock joint surfaces can be calculated using the Barton and Bandis (1990) criterion.

11.5 Pull test

The pull test can also be used to measure the basic friction angle (ϕ_r) and other parameters affecting the shear strength of rock joint surface. The pull test is based on the sliding of a rock block placed on another one. This action is performed by applying a shear force parallel to the contact surface of the two blocks, in order to slide the upper block, under a certain normal stress.

González et al. (2014) proposed a method to perform the pull tests using a simple setup shown in Figure 11.12. In this research, the steps of the pull test were as follows:

1 Specimens are cut to specific dimensions.
2 The lower half of the block is placed horizontally on the test plate and fixed in place. A level bubble is used to check the horizontality of the test plate.

Figure 11.12 Pull test device (González et al. 2014).

3 The weight of the block's upper half is measured and placed on the block's lower half and is hooked to a pulley using a wire tied horizontally around it and connected to a bucket at the end by a grip.
4 Inside the bucket is poured with sand until the upper half of the block is pulled. Then the weight of the sand inside the bucket is measured and recorded.
5 Since small displacements may sometimes occur during full displacement, only tests in which their displacements amount to at least 10% of the specimen length are considered valid.
6 Each experiment is repeated five times, and the maximum and minimum values are removed. The specimen surface is cleaned with a dry cloth or brush before repeating the test, and the specimens and the wire are placed in the same initial position.
7 As a result of the pull test, the shear stress is obtained. Using the normal stress, the basic friction angle (ϕ_r) for each test can be obtained. Finally, the basic friction angle is obtained by averaging the results of the experiments.

One of the important goals of the research of González et al. (2014) was to compare the results obtained from the tilt and the pull tests with the same specimens. According to these researchers, both mentioned tests provide acceptable results for calculating the basic friction angle (ϕ_r) of rock joint surface, although the value of the standard deviation in the pull test is much lower than that of the tilt test. Moreover, there is a good correlation between the results of these two tests.

Fathalizadeh (2022) investigated the basic friction angle of rock joint surfaces in different directions of a specimen surface and the upper block weight effect using a pull test device made in Damghan University. In this research, the pull test was done with three different normal stresses applied on the joint surfaces and also in four specified directions on the specimens, where these three stresses are:

- Normal stress resulting from the weight of the upper half of specimen.
- Normal stress resulting from the weight of the upper half of specimen along with a four-kilogram weight.
- Normal stress resulting from the weight of the upper half of specimen along with an eight-kilogram weight.

In general, to perform the pull test, the lower block of specimen is placed horizontally on the test plate and fixed in place. A level bubble is used to check the horizontality of the test plate. After weighing the upper block of the specimen, this block is placed on the lower block and connected to a spring scale using a wire tied horizontally around it. The spring scale is slowly pulled until the upper block slips on the lower block and displaces. In this test, since it is sometimes possible for small displacements to occur during full displacement, only tests in which the amount of their displacements reaches at least 10% of the specimen length are considered valid. Each test is repeated five times and the maximum and minimum values are removed. The specimen surface is cleaned with a dry cloth or brush before repeating the test, and the specimens and the wire are placed in the same initial position. The performing mechanism of the pull test in the normal stress state resulting from the weight of the specimen's upper half, proposed by Fathalizadeh (2022), is presented in Figure 11.13.

During the pull test, when the upper half of the specimen slips due to the shear force, the force indicated by the scale is recorded. As a result of the pull test, the shear stress can be calculated using the following equation:

$$\tau = \frac{F}{A} \tag{11.15}$$

Figure 11.13 Performing mechanism of the pull test: (a) the lower half of rock block is placed horizontally on the test plate, (b) the upper half of the block is placed on its lower half which is fixed, (c) a wire is tied horizontally around the upper half and connected to a spring scale and (d) the spring scale is slowly pulled until the upper half of the block slides.

In this relation, F is the force shown by the scale at the moment of the specimen's upper block sliding (kN) and A is the contact surface area of the two blocks (area of the rock joint surface in cm^2). The normal stresses (σ_n) resulting from the upper block weight, or the sum of the upper block weight and the weights used on it during the pull test, can be calculated through the following equation:

$$\sigma_n = \frac{W}{A} = \gamma b \tag{11.16}$$

In this equation, W is the specimen's upper block weight (N), A is the contact surface area of the specimen's two halves, γ is the rock dry unit weight (N/m^3) and b is the upper block thickness. Finally, the basic friction angle (ϕ_r) for each test can be obtained through the following equation:

$$\phi_b = \tan^{-1}\left(\frac{\tau}{\sigma_n}\right) = \frac{F}{\gamma b} \tag{11.17}$$

11.6 Punch shear strength

One of the most recent indirect methods for estimating the engineering properties and behavior of intact rock is the use of punch test (PT) results, which has been accepted as a suggested method by the International Society of Rock Mechanics (ISRM) (2007). The first reference regarding the use of the PT can be seen in an article written by Lacharite (1960). In this article, the punch shear test was used to estimate the shear strength of a sandstone, and it was found that the apparent shear strength of the rock increases with increasing specimen diameter. The specimen thickness does not have much effect on the obtained apparent shear strength, except in cases where the specimen is very thin (quoted from Sulukcu and Ulusay 2001). With the help of this method, the shear strength of all rock types can be evaluated in the laboratory and the shear strength of rocks can be compared with their punch shear strength and its performance can be evaluated (Fereidooni 2014). Currently, the PT for rock is done by two methods including block punch test (BPT) and cylindrical punch test (CPT). The indices that can be calculated by these two tests are block punch index (BPI) and cylindrical punch index (CPI), respectively.

11.6.1 Block punch test

The BPT device consists of two main parts including the lower metal part or the main base and the upper metal part or the punching block. Both of these parts should be cut from hard steels that can be resistant against high stress and pressure during the test and not be deformed (Ulusay et al. 2001). Figure 11.14 shows a view of the block punch device made by Ghasemi and Fereidooni (2021) to perform this test. This device consists of a base, punching channel and punching block. In the base section or the lower holder of the device, two narrow steel strips are installed in order to fix the rock specimen inside the device. In this part and in its middle, there is an elongated channel, where the punched part of the specimen is placed inside and separated from the two side parts in the holder. The upper part of the device, which is the punching block, is also designed so that it can be freely placed inside the punching channel.

To perform the BPT, the first required specimens are prepared for the test. For this purpose, disc- or cube-shaped specimens with a thickness between 5 and 15 mm are cut by a diamond saw. In accordance with the ISRM (2007) standard, the surfaces of the specimens should be parallel and their both sides should be smooth and free of severe irregularities. The use of any covering material that prevents direct contact between the upper surface of the specimen and

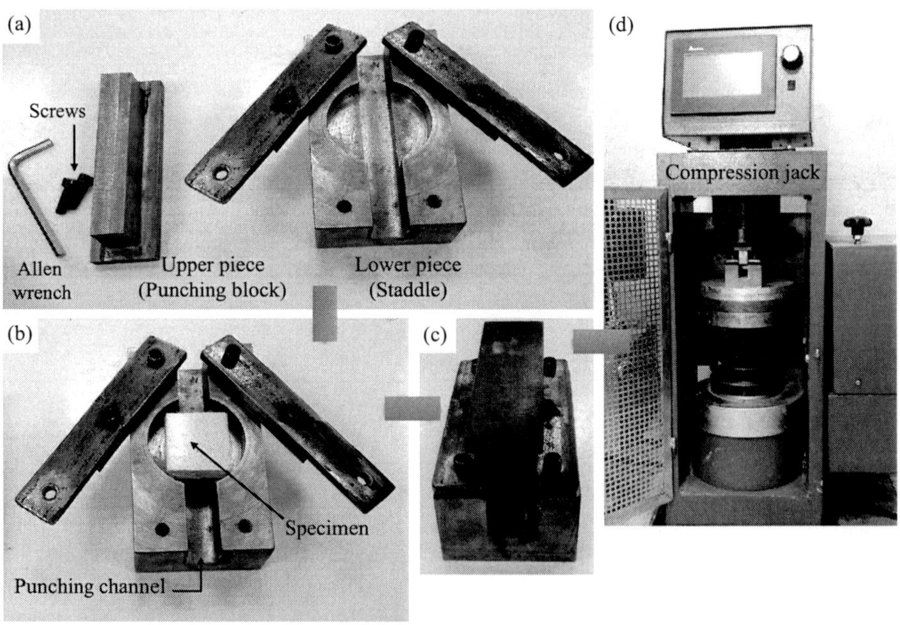

Figure 11.14 An overview of the block punch apparatus and the steps of the test: (a) different parts of the shear device, (b) placing specimen inside the device, (c) the device ready to be placed in the compression jack and (d) placing the block punch device containing the specimen inside the compression jack.

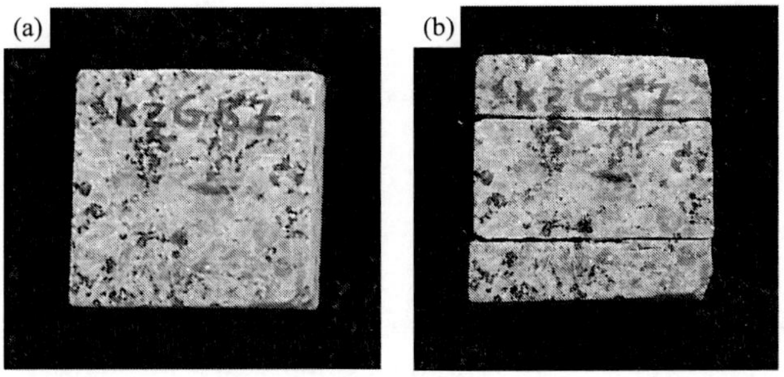

Figure 11.15 An image of the specimen used to perform the block punch test: (a) before and (b) after its failure.

the punching part is not allowed. The diameter and thickness of the specimens are calculated by averaging two values measured in two perpendicular directions with an accuracy of 0.1 mm (Ulusay et al. 2001). After preparation, the specimen is placed inside the jaws of the device and its two side screws are closed to tighten. In the next step, the test device is placed inside a compression jack and then the load is gradually applied to the specimen at a constant rate of 0.05 kN/s until it breaks in a period of 10–60 seconds (Ulusay et al. 2001). The specimen should be loaded axially and vertically.

After the test, the force required to fail the specimen (F) is recorded. After failure, theoretically, the specimen is divided into three parts, two of which are placed on the side and under the steel strips holding the specimen, and the other is inserted into the punch channel. According to the ISRM (2007) standard, the test is unacceptable if the failure surfaces are not completely parallel or not fully developed and cross-fracture surfaces occur. Figure 11.15 shows images of specimens tested by the block punch apparatus before and after failure and their failure pattern. The results of the BPT can be used to estimate the mechanical properties of intact rock, such as compressive strength, tensile strength and shear strength (Khajevand and Fereidooni 2019; Jamshidi and Fereidooni 2022).

To determine the BPI, before the test, the diameter and thickness of the disc-shaped specimen are measured and recorded, and after the test, the force required to fail the measured is read from the apparatus. In this case, the BPI can be calculated through the following equation (Ghasemi and Fereidooni 2019):

$$\mathrm{BPI} = \frac{F}{A} \tag{11.18}$$

In this equation, F is the maximum force applied to specimen in N, which occurs at the moment of failure. A is the area of the shear surface in mm^2. In this case, the value of BPI will be obtained in MPa. According to the thickness (t) and radius (r) of the disc-shaped specimen, the shear surface area can be calculated through the following equation:

$$A = 4t(r^2 - 95.1)^{0.5} \times 10^{-6} \tag{11.19}$$

Van der Schrier (1988) conducted his research using the same device made at Delft University on specimens with a thickness of 10 mm and a diameter of 40 mm and presented a relationship between the uniaxial compressive strength and Brazilian tensile strength with BPI and suggested that BPI is reliable to use as an index to estimate the uniaxial compressive strength and Brazilian tensile strength of rock and provided the following relationships to calculate these parameters:

$$UCS = 6.1BPI - 3.3 \tag{11.20}$$

$$BTS = 0.4BPI - 0.4 \tag{11.21}$$

In all the studies that have been mentioned so far, the effect of specimen size on the BPI has not been considered. But in the following years, comprehensive studies were conducted to investigate the effect of specimen size on BPI by Ulusay and Gokceoglu (1997; 1998; 1999). The results of these studies made it clear that correcting the effect of size in this test is essential and cannot be ignored, and the thickness of 10 mm and the diameter of 50 mm can be relatively considered as standard and suitable dimensions for specimens. These researchers presented new relationships to calculate the modified BPI (BPI_c) as well as the relationship between the uniaxial compressive strength and the modified BPI. The modified block punch strength index (BPI_c or $BPI_{10,50}$) is an index value for a disc-shaped specimen with a thickness of 10 mm and a diameter of 50 mm. For other disc-shaped specimens with any other size, the block punch strength index can be corrected by the constants K_t and K_D, which are called the thickness correction and diameter correction factors, respectively. Therefore, the modified block punch strength index can be calculated in MPa through the following equation:

$$BPI_{10,50} = BPI_c = \frac{F_{10,50}}{A_{10,50}} = \frac{10^{-3} \times F_{t,D} \times K_t \times K_D}{921 \times 10^{-6}} \tag{11.22}$$

In this equation, $F_{t,D}$ is the force required for the punching specimen in kN, and t and D are the thickness and diameter of the specimen in mm, respectively. Thickness and diameter correction coefficients (K_t and K_D) can be obtained using the following relations:

$$K_t = 13.74\, t^{-1.1265} \tag{11.23}$$

$$K_D = 234.53\, D^{-1.3926} \tag{11.24}$$

By inserting the mentioned relationships, finally, the following relationship is obtained to calculate the modified block punch strength index:

$$BPI_c = 3499 D^{-1.3926} \times t^{-1.1265} \times F_{t,D} \tag{11.25}$$

In this equation, D and t are the diameter and thickness of the disc-shaped rock specimen in mm, respectively, and $F_{t,D}$ is the force required to fail the specimen in kN. In this case, the BPI value is in MPa. This relationship was later modified by Sulukcu and Ulusay (2001) based on the results of further tests, and the modified BPI can be calculated according to the following relationship:

$$BPI_c = 3353 D^{-1.3687} \times t^{-1.157} \times F_{t,D} \tag{11.26}$$

In this case, the relationship between uniaxial compressive strength and modified BPI is as follows:

$$\mathrm{UCS} = 5.5\mathrm{BPI}_c \tag{11.27}$$

The value of BPI can be used as a suitable alternative parameter to evaluate the intact rock strength in RMR classification (Fereidooni 2014). Sonmez and Tunusluoglu (2008) presented a relationship for calculating uniaxial compressive strength (UCS) using BPI values, considering the constant m_i of Hoek and Brown's failure criterion. m_i is a dimensionless parameter and depends on the friction angle between rock grains (C_{HB}) and the degree of interlocking of the grains. The rock lithology is involved in the value of the BPI by Hoek and Brown's failure criterion. Using this coefficient when two rocks with different lithologies, which have the same punch block index, to estimate the UCS through the BPI leads to interesting results. These researchers, by using the new Hoek and Brown's failure criterion equation and the data set used in this research, indicated the relationship between these three parameters as follows:

$$\frac{C_{\mathrm{HB}}}{\mathrm{UCS}} = 0.441 m_i^{-0.3824} \tag{11.28}$$

By inserting the equivalent value obtained for C_{HB}, which is equal to 0.8 BPI, the above equation can be presented as the following relationship:

$$\mathrm{UCS} = A \times \mathrm{BPI}_c \tag{11.29}$$

In this equation, the value of A is:

$$A = 0.8 \times 2.266 m_i^{0.3824} \tag{11.30}$$

In this equation, A is the conversion factor, in which the rock lithology is included by Hoek and Brown's failure criterion. The value of m_i is the constant of Hoek and Brown's failure criterion according to the rock type. After calculating the modified block punch strength index, rock classification can be done based on this parameter, as shown in Table 11.1. The modified block punch strength index is about 0.68 time of indirect or Brazilian tensile strength of rock.

Table 11.1 Classification of rocks based on the modified block punch strength index (Sulukcu and Ulusay 2001)

Rock class	*Modified block punch strength index, BPI_c (MPa)*
Very weak	<1
Weak	1–5
Moderate	5–10
Favorable	10–20
Strong	20–50
Very strong	>50

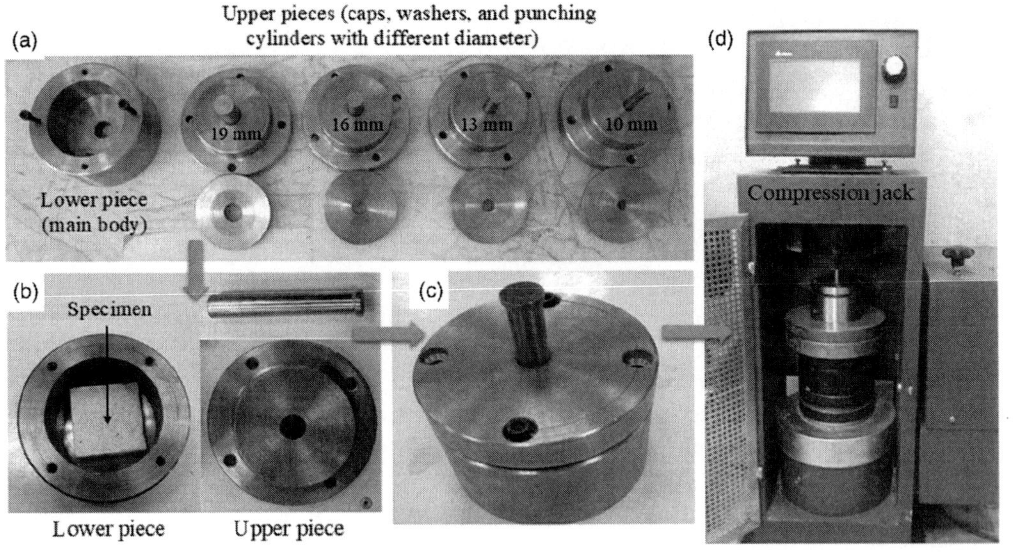

Figure 11.16 An overview of the cylindrical punch apparatus and test mechanism: (a) different parts of the device, (b) placing specimen inside the device, (c) the device ready to be placed inside compression jack and (d) placing the cylindrical punch device containing the specimen inside the compression jack.

11.6.2 Cylindrical punch test

The cylindrical punch device consists of three parts including main body (lower pieces), cap (upper pieces) and punching cylinder, which has the ability to load vertically on specimen and causes a circular failure in the tested rock specimen. The main body has a chamber with a diameter of 80 mm, at the end of which there is a hole of the same diameter as the punching cylinder diameter so that it can move freely inside it. The hole is also present in the cap and is location of the punching cylinder. The cylindrical punch testing device, which is made by Ghasemi and Fereidooni (2021), has punching cylinders in four different diameters of 10, 13, 16- and 19 mm. Figure 11.16 shows a general view of the cylindrical punch apparatus, with its different parts.

To perform the CPT, the required specimens are prepared. The index calculated by this test is called CPI. The cylindrical punch device can also be placed in the uniaxial compressive strength and point load strength testing apparatuses. To perform the test, first, disc- or cube-shaped specimens with dimensions of 54 and thicknesses between 5 and 15 mm, whose two surfaces are parallel to each other, are obtained from rock block. Then the specimen is placed inside the jaws of the device and its two side screws are closed tightly. One of the important points that must be paid attention to when placing the specimen inside the device is the concentricity of the specimens and the bottom hole of the device so that the symmetry of the specimen is maintained during failure. In the next step, the test device is placed inside the uniaxial compressive strength testing apparatus or the point load strength apparatus and is loaded at a rate of about 0.05 kN/s until it breaks between 10 and 60 seconds (Ulusay et al. 2001). In this test, like the BPT, the specimen should be loaded axially and vertically.

After placing the specimen in the compression jack, the force required to fail the specimen is recorded. After performing the CPT, the specimens are divided into several parts, the most

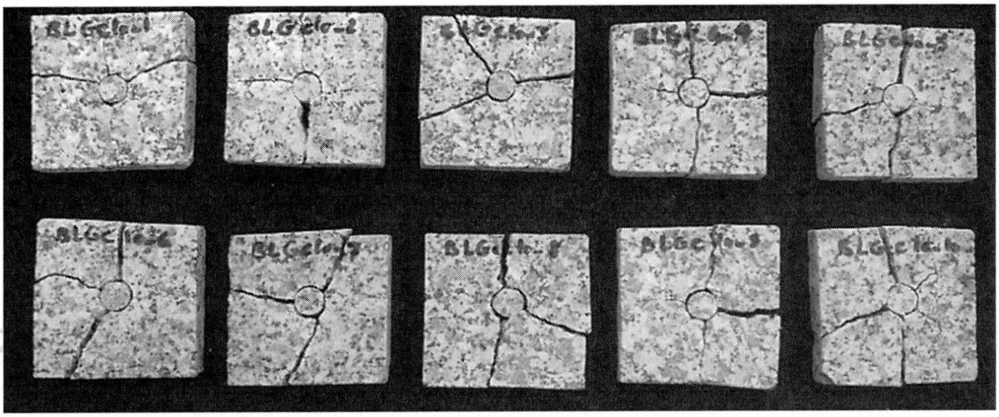

Figure 11.17 Failure pattern of granite specimens failed by the cylindrical punch apparatus.

important part of which is the middle pill-like part, and the parts around it have different sizes of radial fractures. The specimen failure should be such that the cutting cylinder is well formed and its effect can be seen on the two upper and lower surfaces of the specimen. Otherwise, the test result is not acceptable (Ghasemi and Fereidooni 2021). Figure 11.17 shows an image of specimens broken by a cylindrical punch apparatus and their failure pattern.

To determine the CPI, before the test, the diameter and thickness of the specimen are measured, and after the test, the force required to fail the specimen is read from the apparatus. Knowing the diameter of the loading cylinder, the CPI can be calculated through the following equation:

$$\mathrm{CPI} = \frac{P}{A} = \frac{P}{\pi D t} \tag{11.31}$$

In this equation, P is the maximum force applied to the specimen in N, which occurs at the moment of failure, and A is the area of the specimen failure surface in mm^2, D is the diameter of the punching cylinder and t is the thickness of the punched cylinder or specimen in mm. In this case, the CPI value will be obtained in MPa.

The most important advantages of the CPT compared to the BPT are as follows (Fereidooni 2014):

- It requires less force to fail the specimen.
- Shear surface of the specimen is the side surface of a cylinder, and this surface is more uniform than the shear surface of the BPT.
- Distance of the shear cylinder lateral surface to the edge of the tested disc-shaped core is the same, while in the BPT, this distance is not the same, and it is effective on the test results.
- Rotation of the specimen around its central point in the CPT has no effect on the test result, while this is not true in the BPT.
- Stress distribution on the specimen surface in the CPT is more uniform than the BPT.
- Similar to the BPT, there is no need to correct the calculated index (CPI) in the CPT, because the diameter of the punched cylinder is always constant and independent of the specimen diameter.

References

Alejano, L.R., Gonzalez, J., Muralha, J., 2012. Comparison of different techniques of tilt testing and basic friction angle variability assessment. *Rock Mechanics and Rock Engineering*, 45(6), 1023–1035.

Alejano, L.R., Muralha, J., Ulusay, R., Li, C.C., Perez-Rey, I., Karakull, H., Chryssanthakis, P., Aydan, O., 2018. ISRM suggested method for determining the basic friction angle of planar surfaces by means of tilt test. *Rock Mechanics and Rock Engineering*, 51(12), 3853–3859.

ASTM, 1995. *Standard Test Method for Triaxial Compressive Strength of Undrained Rock Core Specimens Without Pore Pressure Measurements. ASTM D2664-95a*. ASTM, West Conshohocken, PA.

ASTM, 2016. *Standard Test Method for Performing Laboratory Direct Shear Strength Tests of Rock Specimens Under Constant Normal Force. Annual Book of ASTM Standards. Designation: D5607-16.* ASTM, West Conshohocken, PA.

Barton, N., 2011. From empiricism, through theory, to problem solving in rock engineering. 6th Leopold Muller Lecture. In: Qian, Q., Zhou, Y., (eds) *Harmonising Rock Engineering and the Environment. 12th ISRM International Congress on Rock Mechanics* (3–14). CRC Press, Beijing.

Barton, N., Bandis, S., 1982. Effects of block size on the the shear behaviour of jointed rock. *The 23rd U.S. Symposium on Rock Mechanics (USRMS)*, American Rock Mechanics Association, Berkeley, CA, pp. 739–760.

Barton, N., 1973. Review of a new shear strength criterion for rock joints. *Engineering Geology*, 7(4), 287–332.

Barton, N., Choubey, V., 1977. The shear strength of rock joints in theory and practice. *Rock Mechanics and Rock Engineering*, 10(1), 1–54.

Barton, N.R., Bandis, S.C., 1990. Review of predictive capabilites of JRC-JCS model in engineering practice. In: Barton, N., Stephansson, O. (eds) *Proceedings of the International Symposium on Rock Joints, held in Loen, Norway* (603–610). Balkema, Rotterdam.

Bruce, I.G., Cruden, D.M., Eaton, T.M., 1989. Use of a tilting table to determine the basic friction angle of hard rock samples. *Canadian Geotechnical*, 26(3), 474–479.

Cawsey, D.C., 1974. The stability of slopes in chalk. PhD thesis. University of Surrey.

Cawsey, D.C., Farrar, N.S., 1976. A simple sliding apparatus for the measurement of rock joint friction. *Geotechnique*, 26(2), 382–386.

Diamantis, K., Fereidooni, D., 2023. Assessing the strength and deformation properties of serpentinite rocks in dry and saturated conditions. *Iranian Journal of Science*, 47(4), 1169–1178.

Fathalizadeh, F., 2022. Assessing the shear strength of rock joint surfaces and its effective parameters using simple laboratory tests. M.Sc. Thesis of engineering geology, Damghan University, Damghan, Iran, 144 pp.

Fereidooni, D., 2004. Investigating the stability of rock slopes along the Ganjnameh-Shahrestaneh road. Master's Thesis in Engineering Geology, Faculty of Science, Bu-Ali Sina Univesity, 192 pp.

Fereidooni, D., 2014. The study of engineering geological properties of metamorphic rocks in Hamedan province with emphasis on anisotropy. PhD Thesis in Engineering Geology, Bu-Ali Sina University, Hamedan, Iran, 224 pp.

Fereidooni, D., 2015. *Principles of Rock Slope Engineering*. Danesh-e-Mana Publications, Hamedan, Iran, 293 pp.

Fereidooni, D., 2018. Influence of discontinuities and clay minerals in their filling materials on the instability of rock slopes. *Geomechanics and Geoengineering*, 13(1), 11–21.

Fereidooni, D., 2023. *Engineering Geology*. First edition, Damghan University Press, Damghan, Iran, 552 pp.

Fereidooni, D., Khajevand, R., (translators), 2019. *Rock Mechanics: An Introduction*. First edition, Damghan University Press, Damghan, 280 pp.

Fereidooni, D., Khanlari, G.R., Heidari, M., 2015. Assessment of a modified rock mass classification system for rock slope stability analysis in the Q-system. *Earth Sciences Research Journal*, 19(2), 147–152.

Ghasemi, F., Fereidooni, D., 2019. Assessing the block and cylindrical punch shear strengths of some building stones used in Iran. *Proceeding of 6th National Conference on Applied Research in Civil Engineering, Architecture and Urban Management*, Tehran, Iran.

Ghasemi, F., Fereidooni, D., 2021. The block and cylindrical punch tests for rocks: A statistical comparison between the results. *Geotechnical Testing Journal*, 45(1), 180–194.

Goldstein, M., Goosev, B., Pvrogovsky, N., Tulinov, R., Turovskaya, A., 1966. Investigation of mechanical properties of cracked rock. *Rock Mechanics and Rock Engineering*, 52(11), 4.

González, J., González-Pastoriza, N., Castro, U., Alejano, L.R., Muralha, J., 2014. Considerations on the laboratory estimate of the basic friction angle of rock joints. *ISRM Regional Symposium-EUROCK 2014*, OnePetro, Sandton, South Africa.

Hoek, E., Bray, J.D., 1974. *Rock Slope Engineering*. Fourth edition. CRC Press, Boca Raton, FL.

ISRM, 1983. Suggested method for determining the strength of rock material in triaxial compression: Revised version. *International Journal of Rock Mechanics and Mining Sciences & Geomechanics Abstracts*, 20(6), 283–290.

ISRM, 2007. Compilation arranged by the ISRM Turkish National Group, Ankara, Turkey. In: Ulusay, R., Hudson, J.A. (eds) *The Blue Book: The Complete ISRM Suggested Methods for Rock Characterization, Testing and Monitoring, 1974–2006*. Kazan Offset Press, Ankara, Turkey.

Jamshidi, A., Fereidooni, D., 2022. Evaluation of the block punch index test for predicting the strength of sandstones. *Scientific Quarterly Journal of Iranian Association of Engineering Geology*, 15(1), 105–118.

JGS, 2008. *Method for Direct Shear Test on a Rock Discontinuity*. Japanese Geotechnical Society, Tokyo, 8 pp.

Khajevand, R., Fereidooni, D., 2019. Utilization of the point load and block punch strengths to predict the mechanical properties of several rock samples using regression analysis methods. *Innovative Infrastructure Solutions*, 4, 1–14.

Kim, D.H., Gratchev, I., Hein, M., Balasubramaniam, A., 2016. The application of normal stress reduction function in tilt tests for different block shapes. *Rock Mechanics and Rock Engineering*, 49(8), 3041–3054.

Lacharite, N., 1960. A study to correlate the shearing, bending and compression properties of rock. Fuels Min Pract Div Mines Branch, Can Dept Mines and Tech Surv, Ottowa, IR 60/38FMIN.17.

Muralha, J., 2012. Rock joint shear tests. Methods, results and relevance for design. *ISRM International Symposium-Eurock 2012*, OnePetro, Sandton, South Africa.

Patton, F.D., 1966. Multiple modes of shear failure in rock and related materials. Thesis, University of Illinois.

Shariati, M., Fereidooni, D., 2021. Rock slope stability evaluation using kinematic and kinetic methods along the Kamyaran-Marivan road, west of Iran. *Journal of Mountain Science*, 18(3), 779–793.

Sonmez, H., Tunusluoglu, C., 2008. New considerations on the use of block punch index for predicting the uniaxial compressive strength of rock material. *International Journal of Rock Mechanics and Mining Sciences*, 45, 1007–1014.

Stimpson, B., 1981. A suggested technique for determining the basic friction angle of rock surfaces using core. *Rock Mechanics and Mining Sciences and Geomechanics Abstracts*, 18(1), 63–65.

Sulukcu, S., Ulusay, R., 2001. Evaluation of the block punch index test with particular reference to the size effect, failure mechanism and its effectiveness in predicting rock strength. *International Journal of Rock Mechanics and Mining Sciences*, 38, 1091–111.

Ulusay, R., Karakul, H., 2016. Assessment of basic friction angles of various rock types from Turkey under dry, wet and submerged conditions and some considerations on tilt testing. *Bulletin of Engineering Geology and the Environment*, 75(4), 1683–1699.

Ulusay, R., Gokceoglu, C., 1997. The modified block punch index test. *Canadian Geotechnical Journal*, 34, 991–1001.

Ulusay, R., Gokceoglu, C., 1998. An experimental study on the size effect in block punch index test. *International Journal of Rock Mechanics and Mining Sciences*, 35(4–5), 628–629 (In: NARMS'98 ISRM International Symposium. Cancun, Mexico, Paper No. 008).

Ulusay, R., Gokceoglu, C., 1999. A new test procedure for the determination of the Block Punch Index and its possible uses in rock engineering. *ISRM News Journal*, 6(1), 50–54.

Ulusay, R., Gokceoglu, C., Sulukcu, S., 2001. Draft ISRM suggested method for determining block punch strength index (BPI). *International Journal of Rock Mechanics and Mining Sciences*, 38, 1113–1119.

Van der Schrier, J.S., 1988. Le «block punch index test». *Bulletin of the International Association of Engineering Geology-Bulletin de l'Association Internationale de Géologie de l'Ingénieur*, 38, 121–126.

Index